Basic Solid-State Electronics

The Configuration and Management of Information Systems

VAN VALKENBURGH,
NOOGER & NEVILLE, INC.

COMMON-CORE

INFORMATION RECEPTION
VOL. 4

AUDIO/VIDEO/DATA/SENSORY RECEPTION
RECEIVING ANTENNAS
AM/FM/COMMUNICATION RECEIVERS
BLACK-AND-WHITE/COLOR TV RECEIVERS
VIDEO RECORDING/DISPLAY TERMINALS
TROUBLESHOOTING/ALIGNMENT

HAYDEN BOOK COMPANY, INC.
Rochelle Park, New Jersey

Library of Congress Cataloging in Publication Data *(Revised)*

Main entry under title:

Basic solid-state electronics.

Includes index.
1. Solid state electronics. I. Van Valkenburgh, Nooger & Neville.
TK7871.85.B3743 1982 621.381 81-13268
ISBN 0-8104-0888-0 (vol. 4)
0-8104-0884-8 (5-vol. ppr.)
0-8104-0890-2 (5-vol. cl.)

Printed in the United States of America

1 2 3 4 5 6 7 8 9 PRINTING

83 84 85 86 87 88 89 90 91 YEAR

ACKNOWLEDGMENTS

Van Valkenburgh, Nooger & Neville, Inc., acknowledges with gratitude the assistance of the following organizations that have supplied large amounts of materials and granted permission for their use in this series. All of these materials come under the heading of *proprietary information*, and were prepared by staff scientists and engineers working at the highest level of their particular specialties. Dedicated people, they contributed greatly to the sum total of data in this series to make it a reservoir of accurate, important, and up-to-date information on solid-state electronics.

Acopian Corp.; American Microsystems, Inc.; Ampex Corp.; Bailey Instrument Corp.; Bell Laboratories; Columbus Instruments; Crown International; Datametrics, Inc.; The Devon-Adair Corp.; Digital Equipment Corp.; Fairchild Semiconductor Division; General Electric Co. — Semiconductor Products Dept.; Globe Union, Inc. — Battery Division; Gulton Industries; Honeywell, Inc. — Solid State Electronics Center; IBM, Inc.; E.F. Johnson Co., LND, Inc.; P.R. Mallory & Co., Inc. — Battery Company; Monroe Calculator Co.; Motorola Semiconductor Products; National Semiconductor Corp.; North American Philips Controls Corp.; RCA Solid-State Division; Signetics Corp.; Union Carbide Corp. — Battery Products Division; and Zenith Radio Corp.

For their assistance, the authors also particularly thank Dr. Anthony P. Uzzo, Director, Radar Systems, Airborne Instruments Laboratory, Eaton Corp.; and Nathan Buitenkant, former Chief Engineer and Editor in charge of the original *Basic Electronics* development.

ELECTROMAGNETIC SPECTRUM
VIDEO
DATA
AUDIO
SENSORY
AUDIO
SENSORY
VIDEO
DATA
OUTPUT MECHANISMS
INFORMATION
RECEPTION

PREFACE

This series is essentially a redevelopment of our earlier work, *Basic Electronics,* originally developed as part of the COMMON CORE® Program —*Basic Electricity, Basic Electronics, Basic Synchros and Servomechanisms,* etc. —for the U.S. Navy during the years 1950-1954. At that time we were concerned with the prerequisite knowledge and skills for the vacuum tube technology of that day as applied primarily to radio communications equipment, radar, and sonar.

Technology wise, although electronics technology over the intervening years has changed *drastically* and *dramatically* via LSI, VLSI, etc., the concepts, purposes, and system building blocks functionally are *essentially the same* — that is, for *communications or information transfer* — with the notable exceptions being the advent of digital/logic switching circuits and their high-density packaging.

Systems wise, the really new aspects have been the internal incorporation of *management* and *control* functions — via digital formatting, processing, storage, retrieval, etc. — into *"intelligent" electronic systems* by means of computers and microprocessors.

Education wise, we faced ironically the same *fractionation* of subject matter as we did back in 1950-1954! Again, there appears to be no one place where a student can go to get a relatively *simple, clear overview* of what electronics is now all about. Again, we have tried to meet this need and challenge by presenting electronics in terms of an *Overall Information Management System.*

Format wise, we continue to use the original, innovative, basic text-format, system-design elements of the COMMON-CORE® Program — the Program by means of which over 100,000 U.S. Navy technicians have been trained along with hundreds of thousands more civilian students and technicians here and in South America, Europe, the Middle East, Asia, Australia, and Africa. This format of proved effectiveness — now incorporating individual learning/testing features and techniques within the texts themselves, and in the accompanying interactive student mastery tests — has withstood the test of time.

This, then, is our second effort in over 25 years to put the basics of electronics back together again by presenting electronics in terms of an *Overall Information Management System,* and in the form of the original innovative basic text-format, system-design elements so that an average person can learn what the basics of electronics — solid-state electronics — are all about.

VAN VALKENBURGH, NOOGER & NEVILLE, INC.

New York, N.Y.

CONTENTS

VOLUME 4 — INFORMATION RECEPTION

Audio/Video/Data/Sensory Reception; Receiving Antennas; AM Receivers; FM Receivers; Communication Receivers; TV Receiver Fundamentals; Black-and-White TV Receivers; Color TV Receivers; Video Recording; Video Display Terminals; Troubleshooting/Alignment

RADIO RECEPTION AND RECEIVER CHARACTERISTICS

INTRODUCTION TO INFORMATION RECEPTION

INTRODUCTION TO RADIO RECEPTION

RADIO RECEIVERS — THEN AND NOW

RADIO CHARACTERISTICS OVERVIEW

RECEIVING ANTENNAS AND AM RECEIVERS

RECEIVING ANTENNAS

AM RECEIVERS

RECEIVING ANTENNAS AND AM RECEIVERS OVERVIEW

FM RECEIVERS

CHARACTERISTICS OF FM RECEIVERS

FM RECEIVERS OVERVIEW

COMMUNICATION RECEIVERS

CHARACTERISTICS OF COMMUNICATION RECEIVERS

COMMUNICATION RECEIVERS OVERVIEW

TELEVISION RECEPTION

TELEVISION RECEPTION FUNDAMENTALS

TELEVISION RECEPTION OVERVIEW

BLACK-AND-WHITE TV RECEIVERS

INTRODUCTION TO BLACK-AND-WHITE TV

BLACK-AND-WHITE TV RECEIVERS OVERVIEW

COLOR TV RECEIVERS AND VIDEO RECORDERS

INTRODUCTION TO COLOR TV/VIDEO RECORDERS

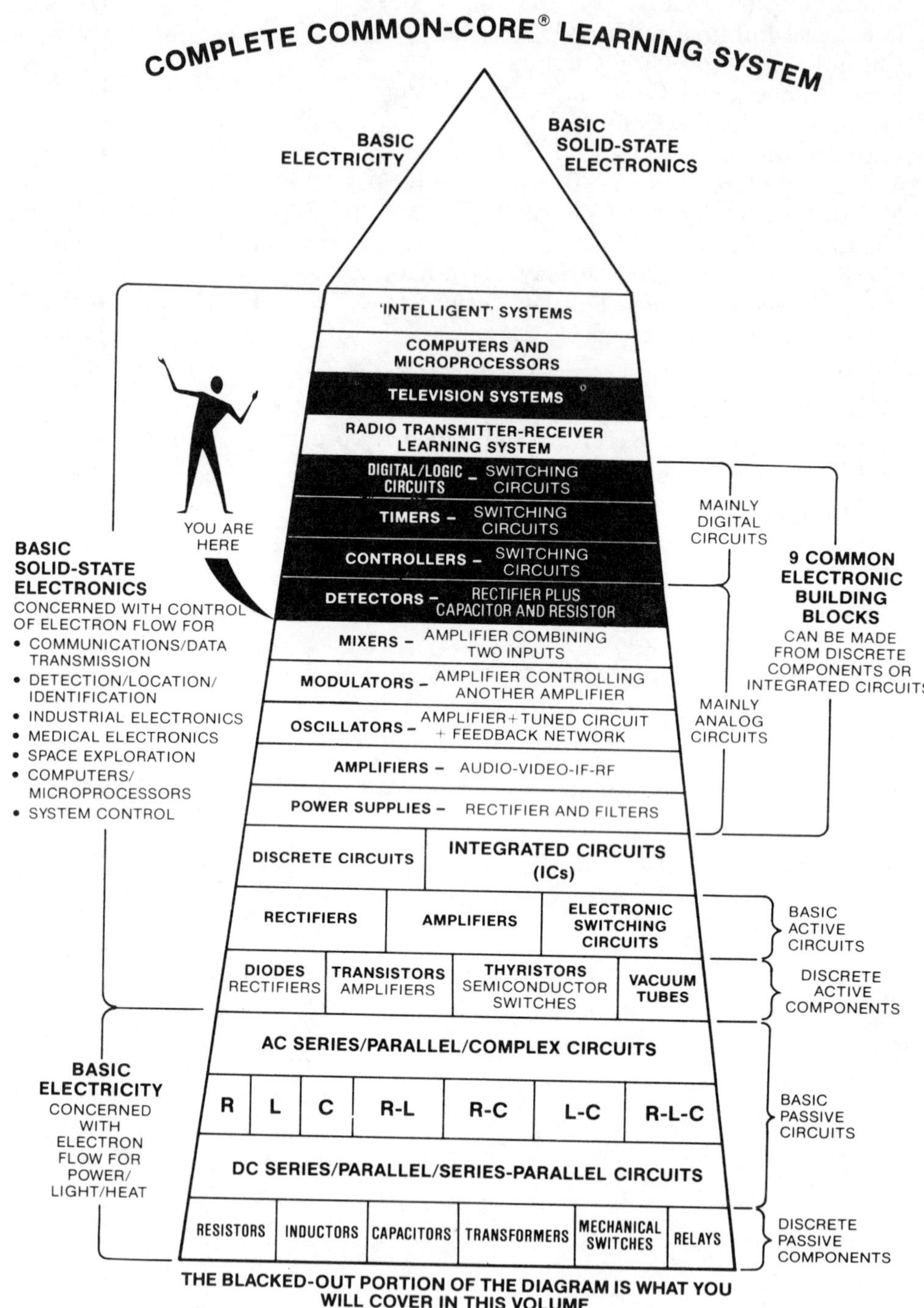

THE BLACKED-OUT PORTION OF THE DIAGRAM IS WHAT YOU WILL COVER IN THIS VOLUME

Information Reception

Information reception is the process whereby an original message—voice, music, picture, alphanumeric data, or sensory information—is recovered from a transmitted signal.

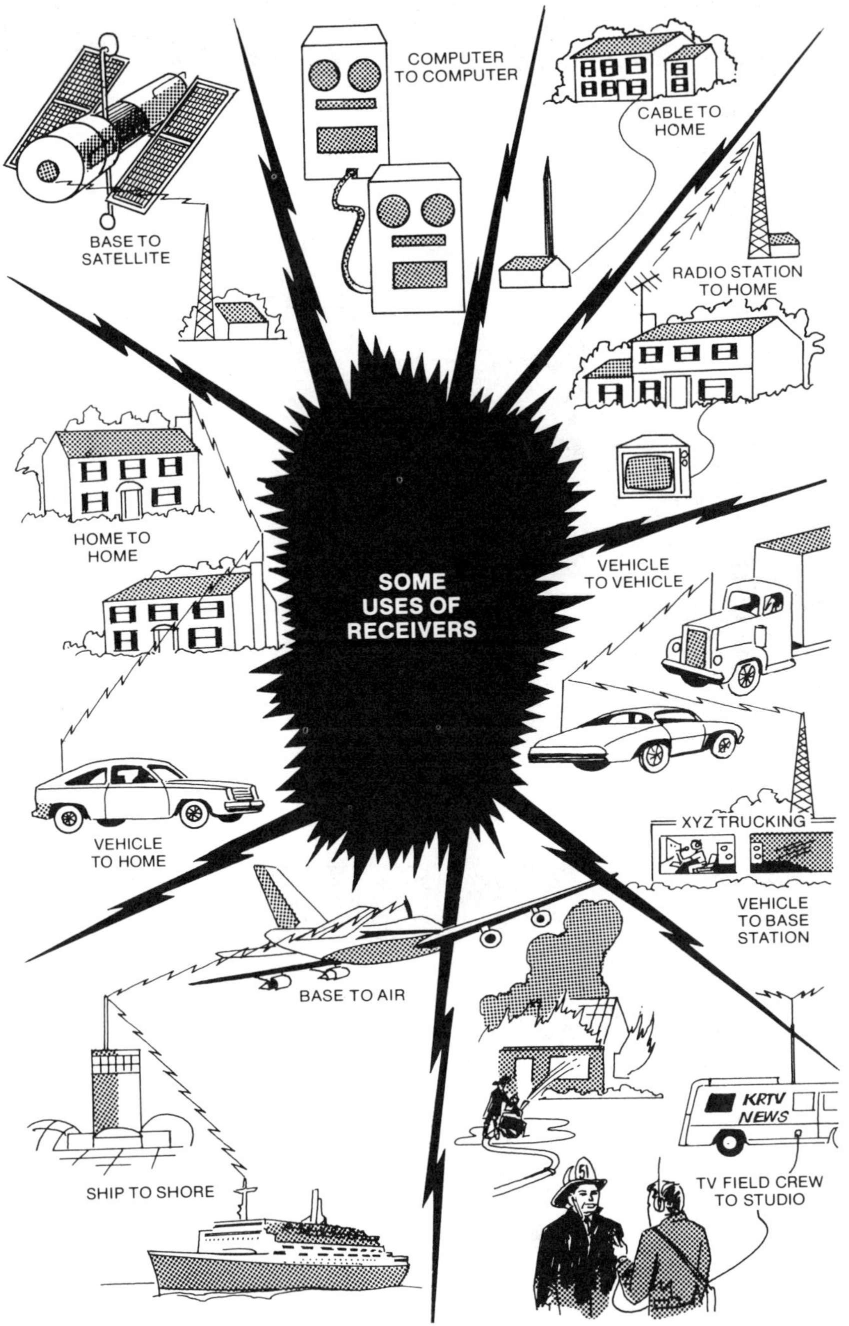

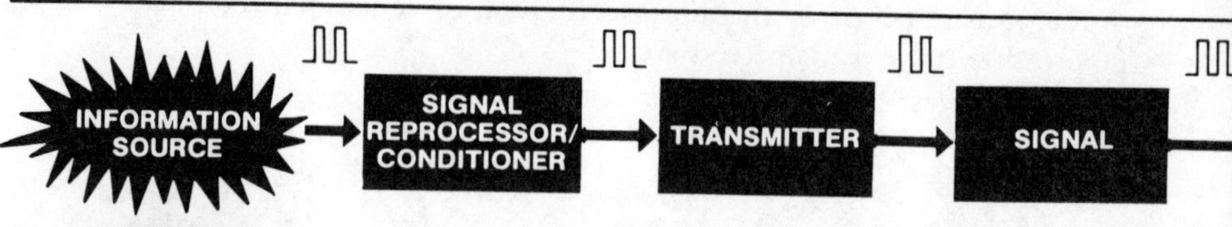

SPEECH
MIND AND CENTRAL NERVOUS SYSTEM
IDEAS/INTELLIGENCE

WORDS INTO CONTINUOUSLY VARYING SOUND PRESSURES TRANSDUCED TO ELECTRICAL SIGNALS

VISION
MIND AND CENTRAL NERVOUS SYSTEM
IDEAS/INTELLIGENCE

PERCEIVED OPTICAL IMAGERY TRANSDUCED INTO CONTINUOUSLY VARYING ELECTRICAL SIGNALS

DATA
STORED INFORMATION-PARALLEL SOURCE
IDEAS/INTELLIGENCE

INFORMATION IN ELECTRONIC MEMORY ENCODED INTO DIGITAL PULSES

TO MACHINE
TO COMPENSATE FOR DISTANCE

INFORMATION TRANSFER SYSTEMS

TELEPHONE
AUDIO
VOICE-WORDS
SENTENCES

VARYING SOUND PRESSURE INTO CONTINUOUSLY VARYING ELECTRIC CURRENT

TELEGRAPH/ TELETYPE/ FACSIMILE/ WORD/DATA PROCESSING DATA
DISCRETE LETTERS
WORDS NUMBERS
IMAGERY

SEQUENCES ENCODED INTO SEQUENCES OF DISCRETE INTERRUPTED ELECTRIC CURRENT

RADIO
AUDIO/DATA
VOICE, MUSIC
DATA INPUT

AUDIO/DATA CONVERTED INTO LINEAR VARYING ELECTRIC CURRENT

TELEVISION
AUDIO/VIDEO
VOICE/MUSIC/
PICTURES/
VISUAL DATA PATTERNS

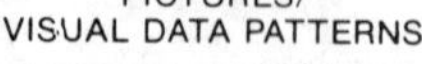

AUDIO/VIDEO CONVERTED INTO CONTINUOUSLY VARYING MULTI-DIMENSIONAL ELECTRIC CURRENT SIGNAL

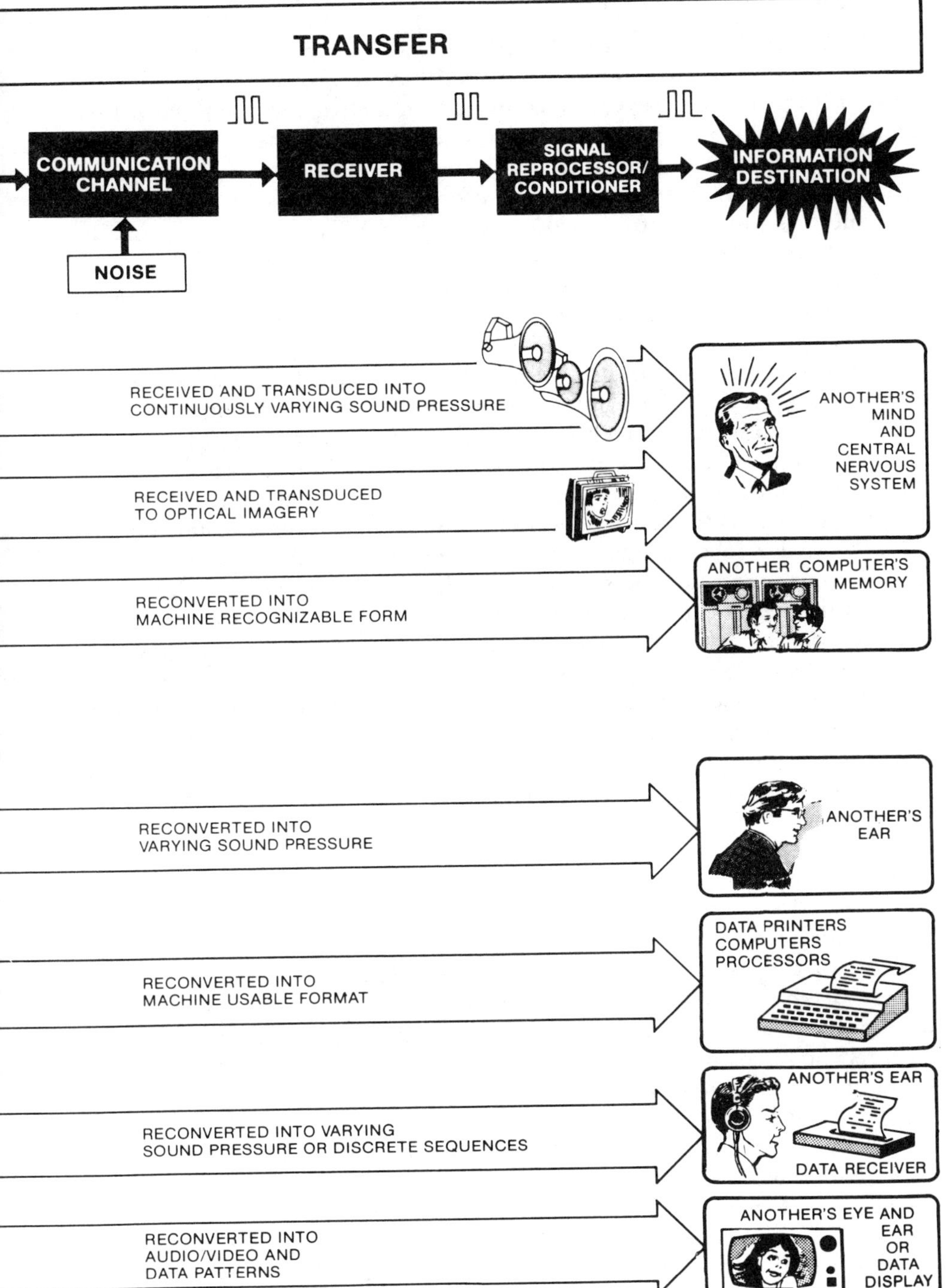
TRANSFER
COMMUNICATION CHANNEL
RECEIVER
SIGNAL REPROCESSOR/ CONDITIONER
INFORMATION DESTINATION
NOISE
RECEIVED AND TRANSDUCED INTO CONTINUOUSLY VARYING SOUND PRESSURE
RECEIVED AND TRANSDUCED TO OPTICAL IMAGERY
ANOTHER'S MIND AND CENTRAL NERVOUS SYSTEM
RECONVERTED INTO MACHINE RECOGNIZABLE FORM
ANOTHER COMPUTER'S MEMORY
RECONVERTED INTO VARYING SOUND PRESSURE
ANOTHER'S EAR
RECONVERTED INTO MACHINE USABLE FORMAT
DATA PRINTERS COMPUTERS PROCESSORS
RECONVERTED INTO VARYING SOUND PRESSURE OR DISCRETE SEQUENCES
ANOTHER'S EAR
DATA RECEIVER
RECONVERTED INTO AUDIO/VIDEO AND DATA PATTERNS
ANOTHER'S EYE AND EAR OR DATA DISPLAY

Information Reception (continued)

The signal may arrive indirectly through propagated radio waves or light beams via an antenna, or directly over cable or wires, or via a fiber optic cable. Whatever the transmission medium and type of information, it is the function of the receiver to isolate the desired signal from any others present and to deliver a *true reproduction* of the original signal to the user. In some cases, the user may elect to have the receiver *filter* and/or store the information to present it in *non-real time*—as in the case of a stock-market quotation or a hotel or airline reservation on a video display terminal. In these cases, the received information is stored and manipulated by computer/microprocessor, then used when called for by the user.

Information reception is also the other half of the information-transfer system, the mirror image of information transmission. This volume is not going to introduce the study of radio and TV receivers, however; you will be encouraged to think of such equipment in broader terms of *information transfer*. We will use radio and radio receivers as a convenient means for a discussion of audio information transfer, just as we will use TV receivers as a convenient means for a discussion of video information transfer. These receivers will lay the foundation for your broader understanding of audio/video/data/sensory information transfer. The broader aspects of control and management of information transfer by computers and microprocessors will be addressed in Volume 5.

The illustrations on the previous two pages show the symmetry of transmission and reception. As you can see, the purpose of the information-transfer system (transmission and reception) is to relay audio, video, data, or sensory information from a source to a point where it is to be used, recorded, or processed with some acceptable standard of quality. The details may vary greatly, but the basic idea is straightforward. Information is transduced into some form of electrical signal that is impressed on a suitable carrier (radio or optical) for propagation to some other point (or many points). Reception starts with the interception of some of the information-bearing carrier. The receiving system processes out the original electrical signal, which, when presented to an appropriate transducer, results in recovery of the desired information.

As you have now progressed over halfway up the COMMON-CORE learning ladder, there is only one new common electronic building block—*detectors*—that you will need to learn about in order to understand the information-receiving system. Actually, a detector is also called a *demodulator*, because it performs the *opposite* function from the modulator you learned about in Volume 3. You will recall that the modulator *impressed* the information *onto* the carrier. A demodulator *recovers* the information *from* the modulated carrier. As you learn about receivers, you will find that, in addition to demodulators, they also use power supplies, amplifiers, oscillators, and mixers—just as these circuits were used to assemble the transmitter portion of the information-transmission system; it is only demodulators that are unique to receivers.

RF Spectrum

In your study of transmitters in Volume 3 you learned that frequencies from subaudio to visible and ultraviolet were used in information-transmission systems. You also learned that broadcasting in most of the spectrum is, of necessity, strictly regulated by United States and international governmental agreements. These regulations do not apply to the noncommercial interception and use of broadcast information. Anyone can build, or buy, the receiving system necessary to monitor radio broadcasts, TV transmissions, satellite communications, aeronautical radio transmissions, or even radar or sonar. The only limitation is that the information cannot be sold or rebroadcast. Also, the receiver must not accidentally radiate signals from its own antenna. In this volume you will study in detail systems ranging in frequency from the AM broadcast band to the UHF TV band. The important thing to remember is that the reception process is for the recovery of the information transmitted in the original signal with *as great fidelity* as is necessary for the intended use.

The electromagnetic spectrum, presented first in Volumes 2 and 3, is shown on the following page for your convenience. A table of standard bands, also presented first in Volumes 2 and 3, is shown below. You will find, in the interests of spectrum conservation, that the signal bandwidth is held to a minimum consistent with the information-transfer requirements. For example, the RF bandwidth of an FM broadcast station is 200 kHz and for TV transmission up to 6 MHz. These bandwidths are not available at lower frequencies, so these transmission/reception systems are confined to VHF and UHF. On the other hand AM broadcasting, short-wave communications, and similar applications require relatively narrow bandwidths (±5 kHz) and can effectively use the lower short- and medium-wave bands. The properties of the ionosphere can strongly influence the range of operation of a system in some frequency ranges. This, in turn, has effects on frequency allocation. Thus, as you can see, the management of the RF spectrum—its use and allocation—is a very complicated matter.

STANDARD BANDS

WAVELENGTH RANGE	BAND NO.	FREQUENCY RANGE	BAND
10^7 – 10^6 m	2	30 TO 300 Hz	ELF (EXTREMELY LOW FREQUENCY)
10^6 m – 100 km	3	300 TO 3,000 Hz	VF (VOICE FREQUENCY)
100 km – 10 km	4	3 TO 30 kHz	VLF (VERY LOW FREQUENCY)
10 km – 1 km	5	30 TO 300 kHz	LF (LOW FREQUENCY)
1 km – 100 m	6	300 TO 3,000 kHz	MF (MEDIUM FREQUENCY)
100 m – 10 m	7	3 TO 30 MHz	HF (HIGH FREQUENCY)
10 m – 1 m	8	30 TO 300 MHz	VHF (VERY HIGH FREQUENCY)
1 m – 10 cm	9	300 TO 3,000 MHz	UHF (ULTRAHIGH FREQUENCY)
10 cm – 1 cm	10	3 TO 30 GHz	SHF (SUPERHIGH FREQUENCY)
10 mm – 1 mm	11	30 TO 300 GHz	EHF (EXTREMELY HIGH FREQUENCY)
1 mm – 100 μm	12	300 TO 3,000 GHz	LONG-WAVE INFRARED
100 μm – 10 μm	13	0.3×10^{13} TO 3×10^{13} Hz	FAR INFRARED
10 μm – 1 μm	14	0.3×10^{14} TO 3×10^{14} Hz	NEAR INFRARED
1 μm – 100 nm	15	0.3×10^{15} TO 3×10^{15} Hz	ULTRAVIOLET AND VISIBLE LIGHT

ELECTROMAGNETIC FREQUENCY SPECTRUM WITH ASSOCIATED EQUIPMENT SYSTEMS

WAVELENGTH OF METERS → 10^{-7}, 10^{-6}, 10^{-5}, 10^{-4}, 10^{-3}, 10^{-2}, 10^{-1}, 10^{0}, 10^{1}, 10^{2}, 10^{3}, 10^{4}, 10^{5}, 10^{6}, 10^{7}

HERTZ → 10^{15}, 10^{14}, 10^{13}, 10^{12}, 10^{11}, 10^{10}, 10^{9}, 10^{8}, 10^{7}, 10^{6}, 10^{5}, 10^{4}, 10^{3}, 10^{2}

Bands: UV, VL, INFRARED, EHF, SHF, UHF, VHF, HF, MF, LF, VLF, VF, ELF

Band boundaries: 3×10^{15}, 3×10^{14}, 3×10^{11}, 3×10^{10}, 3×10^{9}, 3×10^{8}, 3×10^{7}, 3×10^{6}, 3×10^{5}, 3×10^{4}, 3×10^{3}, 3×10^{2}, 3×10^{1}

MILLIMETER WAVES

MICROWAVES

LINE-OF-SIGHT TRANSMISSION

GROUND AND SKY WAVES

SHORT WAVES

AUDIO FREQUENCIES

(14) SPECIAL EQUIPMENT

(13) ELECTRO-OPTICAL DEVICES
COMMUNICATIONS

(12) COMMUNICATIONS
RADAR
SPECIAL DETECTION
ELECTRO-OPTICAL
DATA LINKS

(11) RADAR
COMMUNICATIONS
SPECIALIZED EQUIPMENT

(10) RADAR
COMMUNICATIONS
SPECIALIZED EQUIP

(9) TV • COMMUNICATIONS
DATALINKS • RADAR
SATELLITE COMMUN

(8) TV — FM
COMMUNICATIONS
NAVIGATIONAL
AIDS • DATA LINKS

(7) SHORT-WAVE BROADCAST
AMATEUR BROADCAST
COMMUNICATIONS
CITIZENS BAND

(6) U.S. STANDARD BROADCAST
COMMUNICATIONS

(5) AERONAUTICAL RADIO
NAVIGATION

(4) MARITIME MOBILE RADIO
NAVIGATIONAL AIDS
LONG-RANGE COMMUNICATION

(3) (2) (1) HIGH-FIDELITY SOUND SYSTEMS

Radio Transmitter-Receiver System

In Volume 1 of this series, you were introduced to the block diagram of a complete radio transmitter-receiver system and learned that this would be the learning *vehicle* which would take you from basic concepts and circuits to a working knowledge of complete radio transmitter-receiver systems.

The arrangement is typical of that used when a radio station transmits a program to a radio receiver in your home. It likewise applies to the arrangement used when one amateur radio station transmits a message to another or when a Citizens Band (CB) operator in one automobile talks to another operator several miles away. Although these various systems differ in size, cost, operating range, and signal quality, they all use the basic arrangement shown here.

In previous volumes you learned about power supplies; audio, video, IF, and RF amplifiers; modulators; oscillators; mixers; and transducers—all the basic electronic building blocks that are needed to make up complete radio transmitters and receivers. Now that you are about to enter the world of receivers it is appropriate to review the operation of the transmitter-receiver system, to firmly establish a frame of reference on which to base the details of this volume.

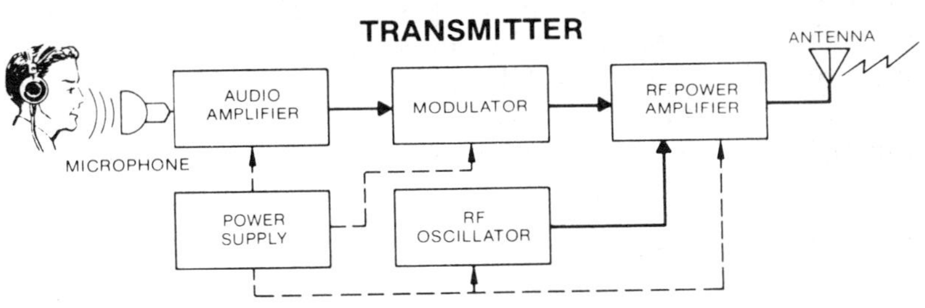

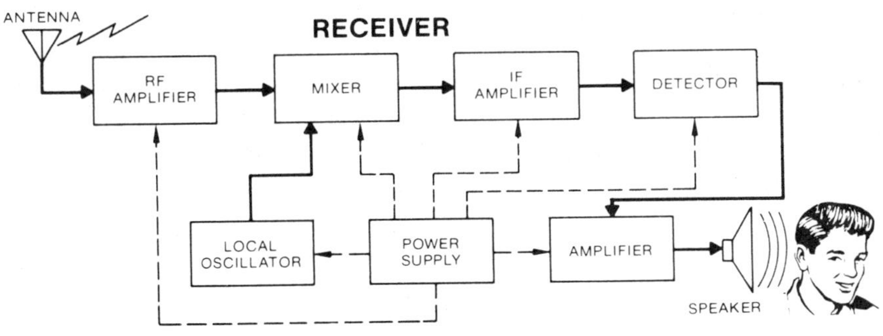

Summary of Transmitter-Receiver Operation

In your study of transmitters in Volume 3, you learned that although the electrical signals (analogs) that duplicate sound waves travel very easily through a conducting wire, they are incapable of radiating readily through air and space to a distant point. For electrical signals to radiate conveniently into space without a wire, their *frequency* must be *considerably higher*. Because of this, it is necessary to generate a high-frequency signal known as a *carrier*, and to superimpose the information on this signal by means of a process known as *modulation*.

In the AM transmitter, a fixed radio-frequency (RF) carrier wave is generated by the *oscillator*. A *microphone* is used to convert the acoustical pressure of the voice message to an electrical signal, and an *audio amplifier* steps up this signal to a usable voltage level. Both the voice and constant-level RF carrier signals are fed into the *modulated* RF amplifier. In an AM system, the signal changes the *amplitude* of the carrier in the modulated RF stage so that the resulting peaks and valleys duplicate the variations in the input signal. Generally this is done by modulating the final power amplifier; it can, however, be done at low level if the subsequent power amplifiers are linear in operation. The output signal is then fed to a suitable antenna system. The block diagram of the AM transmitter operation—presented first in Volume 1—is shown at the bottom of this page for your convenience.

In an FM system, the audio signal varies the carrier *frequency* of the oscillator by an amount equal to the audio signal amplitude, and at a rate equal to the signal frequency. This frequency-modulated RF carrier is fed into a *power amplifier*, which steps it up to a suitable power level. This output signal is fed to the *antenna* to be radiated.

AM TRANSMITTER OPERATION

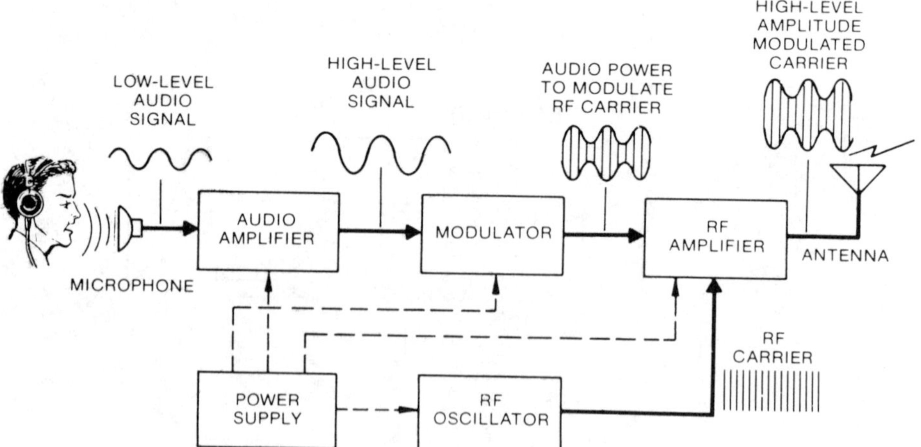

Summary of Transmitter-Receiver Operation (continued)

At the receiver it is necessary to remove the carrier signal from the received, modulated RF signal and leave only the electrical equivalent of the original information. The problem is that the received signal is quite weak, often only a few millionths of a volt. Furthermore, selective circuits are necessary so that the desired signal can be picked out from among all the various signals present from other transmitters on nearby frequencies. The weak signal, picked up by the receiver *antenna*, is first stepped up and preselected by an *RF amplifier* to raise it to a level suitable for signal processing to begin. Because amplification and signal selection at the carrier frequency are not always convenient, the modulated carrier is usually converted to a lower *intermediate frequency* (IF), which is independent of the incoming carrier frequency. This conversion is accomplished by the *mixer*, which takes in both the modulated RF carrier and a separate local oscillator signal. The oscillator frequency differs from the received carrier by a fixed *intermediate frequency* (IF). The variations in the mixer output signal duplicate those of the original received signal but at the lower frequency of the IF carrier. Considerable amplification can now be accomplished by the narrow-band fixed-tuned *IF amplifier.* The *detector* (or *demodulator*) removes the intermediate-frequency component from the output of the IF amplifier, and only an electrical duplicate of the original voice signal remains. Now an *audio amplifier* boosts this signal to a level sufficient for the *speaker* to convert it back to sound duplicating the original voice message.

The pages that follow will review the outstanding features of radio receivers in much greater detail. Stress will be placed on the organization of these systems in terms of the basic building blocks and how these circuits interconnect and work together.

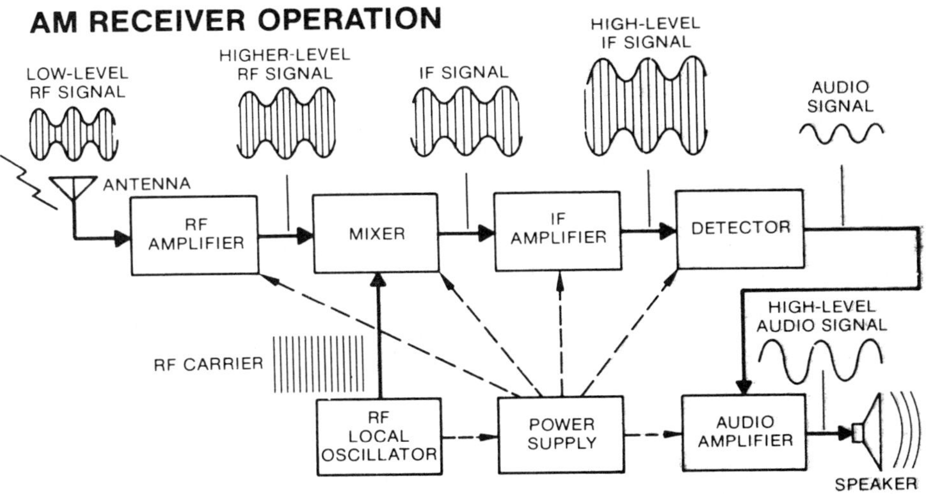

Modes of Reception—AM

The mode of receiver operation is determined by the method *originally used* to impress information onto the continuous-wave (CW) carrier signal. In Volume 3 you learned about the various types of modulation.

As you have learned, in the simplest method of AM, A1 (telegraphy), the carrier is interrupted, or simply turned on and off. The result is the familiar dot-dash or dit-dah signal. This method is used primarily by amateur radio and special long-distance communications systems. If the receiver is tuned to exactly the same frequency as the transmitter, the intelligence appears as a level shift in the output, with no audible signal present. If an audio output is desired, a beat-frequency oscillator (BFO) is used. The BFO is tuned to a frequency a few hundred hertz from the IF and mixed with it. This produces an audio tone at the receiver output when the transmitter is on.

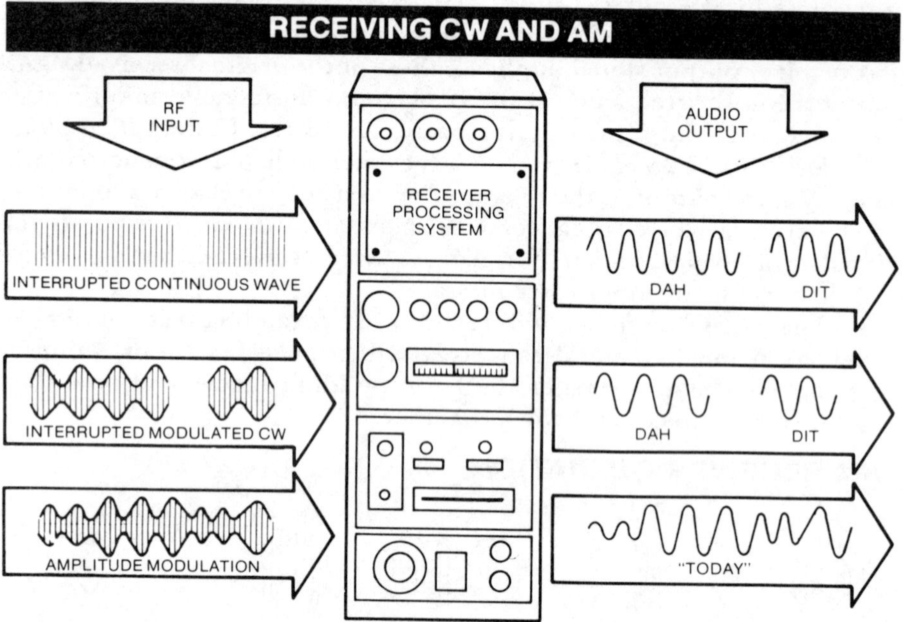

A2 telegraphy (interrupted modulated CW) uses fixed-frequency amplitude modulation of the carrier to provide an audible tone at any AM receiver output without the use of a BFO.

A3 telephony, as you will recall from your study of transmitters, is normal amplitude modulation in which the information is in the double sidebands of the RF carrier. For a reduced-carrier system, the modulation is designated as A3B; for a single-sideband transmission with a reduced carrier, the modulation is designated A3J.

Since AM causes the amplitude of the carrier signal to vary in accordance with the information being transmitted, and AM receiver must be configured to *detect* or *demodulate* these amplitude variations to recover the original information.

Modes of Reception—FM

In FM transmission, the information signal causes the frequency rather than the amplitude, of the RF carrier to change. F1 (telegraphy) uses frequency-shift keying (FSK) to cause the carrier to shift between two frequencies, with one frequency representing the carrier *on* and the other frequency representing the carrier *off* (analogous to A1 modulation). F2 uses a tone, as in A2 modulation, except that frequency—not amplitude—is varied by the tone. F3 telephony is used in FM broadcasting (including stereo) and in communication systems. In telephony, the information-signal frequency controls the *rate* at which the RF carrier frequency is changed, and the information-signal *amplitude* controls the amount by which the carrier frequency is changed—called its frequency *deviation.*

Reception of FM requires a method of detecting the frequency variation of the RF carrier. In your study of FM and TV receivers in this volume you will learn several different methods of detecting FM. The FM demodulator or detector is often called a *discriminator.* Below is a figure describing FM, TV, and FAX reception.

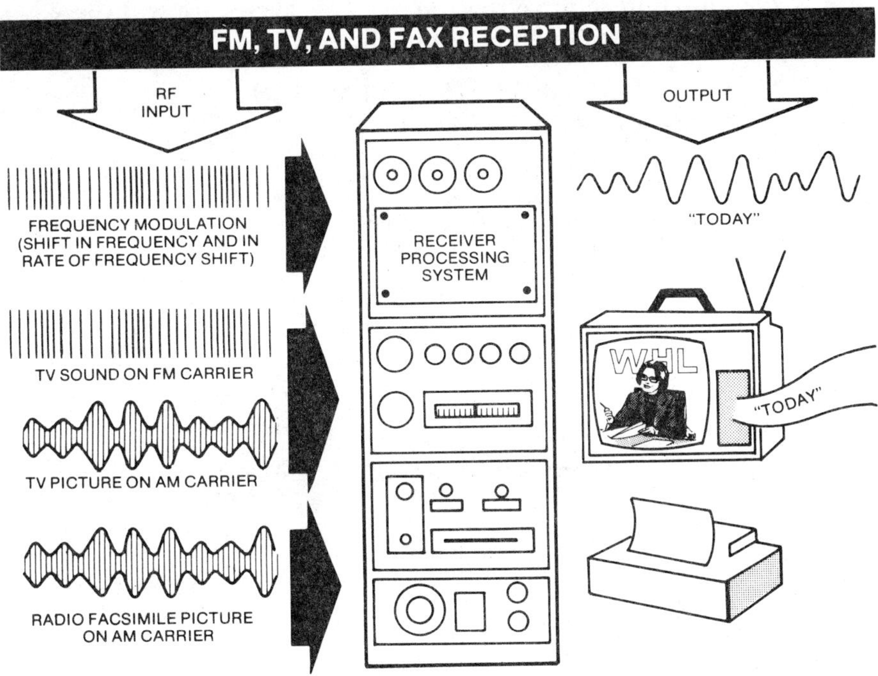

Regardless of the method of modulation used, the receiver has a very small signal at its input. This signal must be amplified and then demodulated to reproduce the information originally sent by the transmission system. This volume will describe the techniques used in modern receivers to enable them to retrieve the information originally sent, and will cover AM and FM reception, communication, and television (both color and black-and-white) systems.

The Diode Detector

Detection, or demodulation, is the process of recovering from an RF carrier the information that was originally impressed on it. For AM, detection means recovery, from the modulated wave, of a voltage that varies in accordance with the carrier amplitude. In most cases, this recovery is accomplished by rectification of the modulated wave, using a diode. The diode operates as a detector in the same way it operated as a rectifier, and you will note that the diode AM detector is almost identical to the half-wave rectifier/filter that you studied in Volume 1. For a steady carrier wave with no modulation, a dc voltage is produced. For an AM wave, the carrier wave *varies*, and hence the output is a *varying dc* or, in our case, the *envelope* of the carrier, which is the desired information.

Consider the simple diode rectifier with load R_L. When an unmodulated carrier signal is applied to the detector, the unfiltered output across R_L consists of half cycles of the input carrier and is thus pulsating dc. When an amplitude-modulated carrier is applied, the output consists of RF carrier pulses whose average (dc) amplitude varies in accordance with the modulation. Notice that the RF half cycles are still there, and the average value is well below peak value.

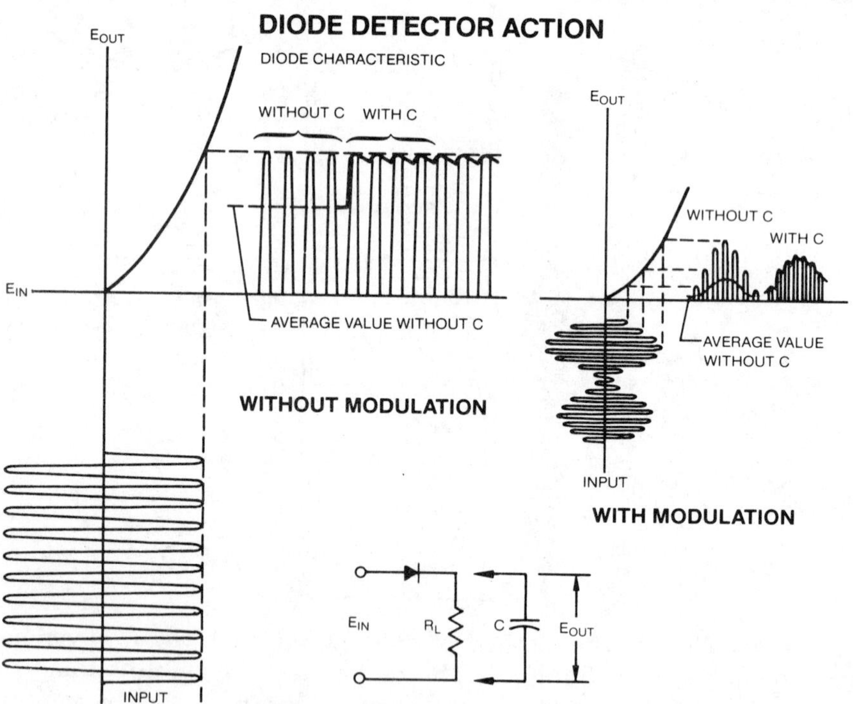

When a capacitor (C) is placed across R_L, the pulsations are smoothed out, just as in a capacitor input filter for a power supply. Now, output is steady dc for a steady carrier and a varying dc representing the modulation signal, but now the modulation signal is near peak value.

The Diode Detector (continued)

The capacitor value is chosen so that there is little discharge between the carrier pulses, but the capacitor can charge and discharge to follow the modulating signal.

In practice, the AM detector circuit usually contains additional RF carrier filtering (R1 and C1) to remove any components of the pulsating carrier wave produced by the diode. In addition, the output from the diode has an average dc value (taken over a long term of several tenths of a second) that is proportional to the average carrier amplitude. Thus, two signals are available, by suitable filtering, from the detector output—one representing the variation of the carrier due to the modulated carrier input and the other representing the average carrier level.

FILTER PASSES ONLY DC TO MODULATION FREQUENCIES

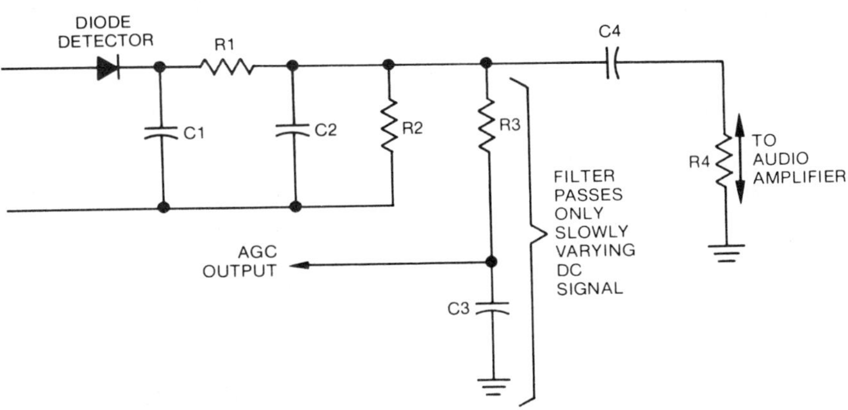

The diagram shows a practical diode detector for receiving AM. The diode detector produces the unidirectional carrier pulses, as described previously. Capacitors C1 and C2 and resistors R1 and R2 act as a low-pass filter to pass signals ranging from dc through the highest modulation frequency of interest. Because of the need for the filter to recover the modulation, the carrier frequency has to be many times the modulating frequency. This signal output is passed to the volume control R4 via the dc-blocking capacitor C4 so that only the desired audio information is sent to the audio amplifier.

Resistor R3 and capacitor C3 function as a low-pass, long-time-constant filter that removes the audio signal, leaving only a slowly varying dc signal that represents the average carrier level. This signal can be fed back to the IF amplifier and used to control the input for some applications to hold the carrier at the detector constant in level (to be discussed later). This is called *automatic gain control (AGC)* voltage.

As you can see, the diode detector functions to strip the modulation signal from the carrier to provide the audio signal. To do this, a filter is used at the detector output to filter the carrier but still allow the variations due to modulation to come through.

Crystal Sets

To aid you in understanding modern radio receivers and receiving circuits, it is helpful to review some of the circuits and receivers that were developed during the evolution from the earliest AM radio receivers up to the modern AM/FM radio sets used today. Most of these early circuits are still used in modern radios, with solid-state devices replacing their vacuum tube counterparts.

The first receivers were used in the early 1900s and were called *crystal sets*. In those days, AM only was used for broadcasting. In its simplest form, the crystal set consisted of an antenna, a tuned circuit, a crystal detector, and a pair of earphones. The crystal detector operated on the principle that certain materials (for example, the mineral galena) would act as a detector if a sharpened fine wire (called the catwhisker), was allowed to rest lightly against a sensitive spot on a chunk of the mineral. Actually, the junction between the galena and the catwhisker was a *semiconductor point-contact junction*, which you learned about in Volume 1, and it *functioned as a diode*. Thus, such a device is a *detector*.

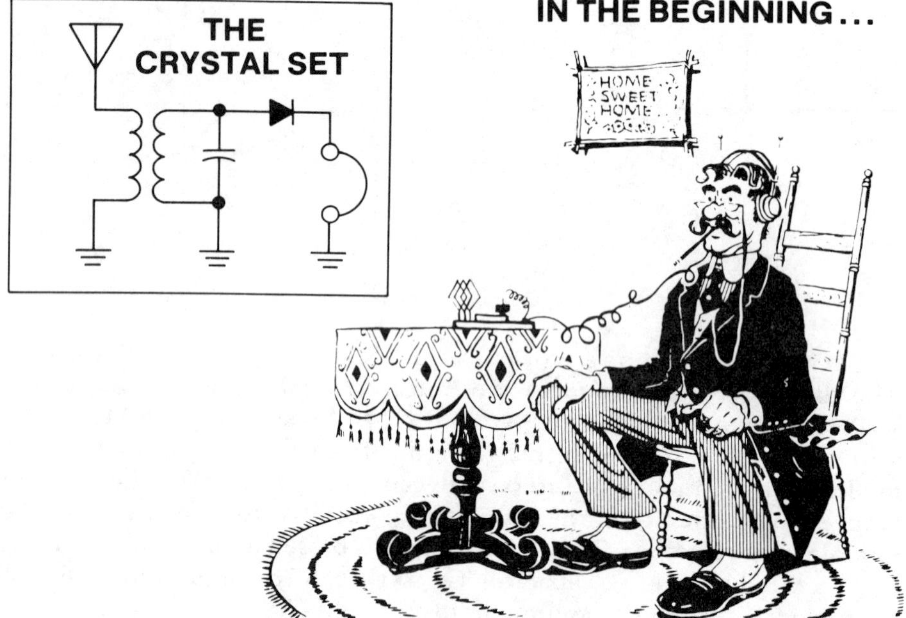

The antenna picked up any signals in the air—in those days there were very few—and the crystal (a primitive diode rectifier) allowed the antenna currents to flow through the headphones on one half cycle of the RF, but blocked the other half cycle. These pulses of RF were filtered in the headphones to make them operate so as to produce an output that was proportional to the modulating wave. Crystal sets at best had one tuned circuit before the crystal, and therefore the selectivity was very poor. Because no amplifiers were used, sensitivity was so bad that crystal sets could be used only for receiving strong local stations. Today these sets have only limited practical application.

Tuned Radio-Frequency Receiver

When the vacuum tube became available, crystal sets were replaced by *tuned radio-frequency (TRF)* receivers that use vacuum tubes as RF amplifiers to provide better selectivity and sensitivity than the old crystal sets could. A vacuum-tube detector does the same thing as the crystal detector. After the detector, the audio signal is amplified in an audio amplifier. The output of the audio amplifier can be a powerful signal, which can be used to drive a loudspeaker or several pairs of earphones.

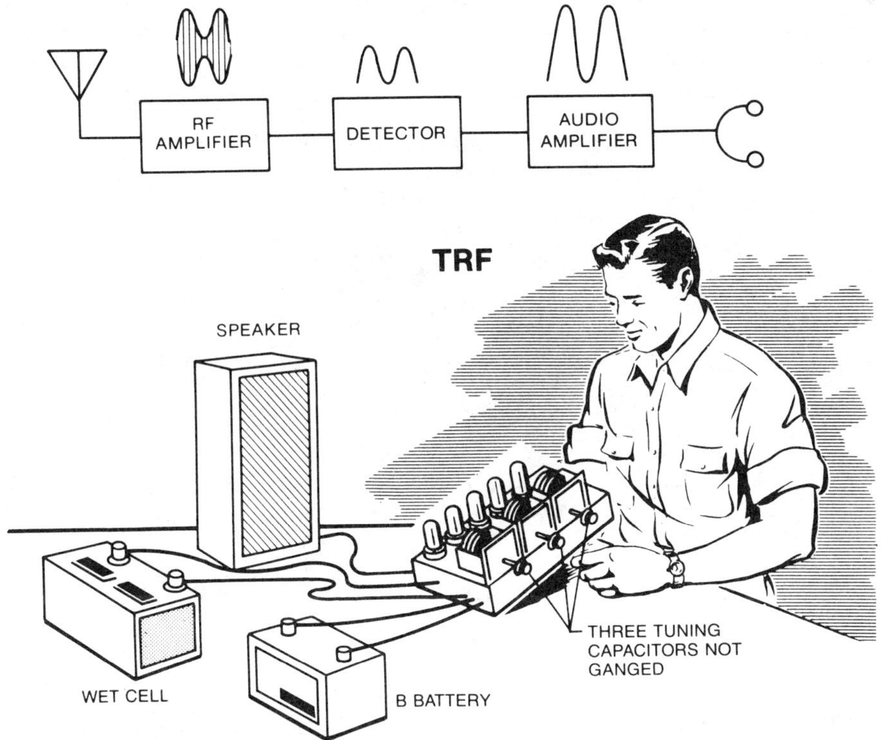

To increase sensitivity, the TRF receiver contains one or more stages of RF amplification *ahead* of the detector. The main purpose of these amplifiers is to provide *additional* selectivity and sensitivity. *Selectivity* indicates how well a receiver receives a desired signal and also rejects unwanted signals; *sensitivity* is a measure of the receiver's ability to pick up a weak signal. (Selectivity and sensitivity will be discussed in greater detail later in this volume.) In general, the more RF amplifier stages used, in TRF receivers, the greater the selectivity and sensitivity will be. On this and the following few pages you will review some of the outstanding points about RF amplifiers. Since the RF amplifier stage is designed primarily for voltage amplification, any transistor or IC suitable for voltage amplification may be used, provided the component has the necessary frequency-response capability to provide amplification at the desired frequency. The RF amplifiers discussed here are the same as those you learned about in Volume 2.

TRF Receivers/RF Amplifiers

TRF receivers have a minimum of two, and usually more, tuned circuits, one associated with each RF amplifier and one with the detector. In the early days of the TRF receiver, each variable capacitor was coupled to its own individual tuning knob. To tune your radio to a station, you had to adjust each knob individually until each tuned circuit was resonant to the frequency of the desired station.

The need for individual tuning knobs was eliminated by having the variable capacitors of all tuned circuits mounted on *one shaft*. This is called *ganged tuning*. Since all capacitors are varied together, then all tuned circuits should be resonant to the same frequency at the same time—making it easier to tune for maximum sensitivity and selectivity.

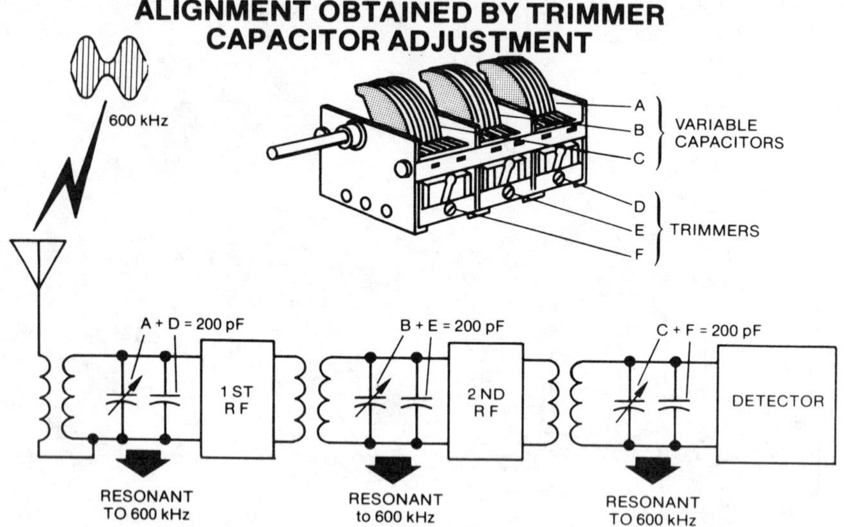

Unfortunately, no two capacitors can be manufactured *exactly alike*; neither can two individual circuits be exactly alike. If nothing were done to compensate for these differences, the tuned circuits in a receiver would be resonant to slightly different frequencies—resulting in reduced receiver selectivity and sensitivity. Such a receiver is said to be *out of alignment*. The problem of misalignment is solved by adding small variable capacitors, called *trimmer capacitors*, in parallel with the main variable-tuning capacitors. Sometimes an adjustment is made in the coil of the tuned circuit rather than on the capacitors. Here, an iron-cored slug is moved in and out of the coil, causing inductance to vary (slug tuning).

In receivers covering only one band, trimmers are usually located on the ganged capacitors, one for each section. In receivers using band switching, additional trimmers for each range are usually mounted on, and in parallel with, the individual coils or the individual coils are adjustable. Trimming is usually done at the high end of the dial since this is usually where error in tuned circuits due to components will have the most serious effect on selectivity and sensitivity.

Experiment/Application—A Crystal Receiver

You can build a simple receiver that will allow detection of local AM radio stations. Only a detector, tuned circuit, and pair of headphones (transducer) are necessary. Suppose you hooked up the circuit shown.

SIMPLE CRYSTAL RECEIVER

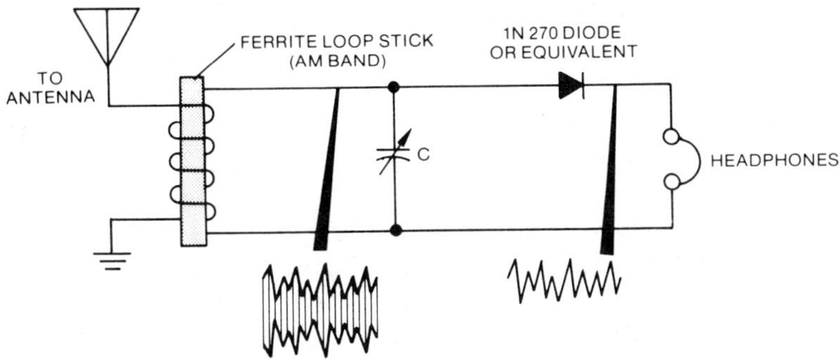

With this circuit, you can receive broadcasts in the AM band for local powerful stations. You must choose the capacitor and coil so that they tune over the AM band. These components are available from any radio-parts distributor. Once the receiver is built, you can observe the signal before and after detection by using an oscilloscope. If you do this, you will notice that the detector acts as you would expect—that is, it takes the intelligence-modulated carrier and strips off the information. You will need a good antenna to get reasonable results because there is no amplification in a simple crystal detector.

You can improve the performance markedly by adding an RF amplifier stage. You can build a simple RF amplifier with a 2N2222 transistor and power supply. Suppose you hooked up the circuit shown below.

RF AMPLIFIER

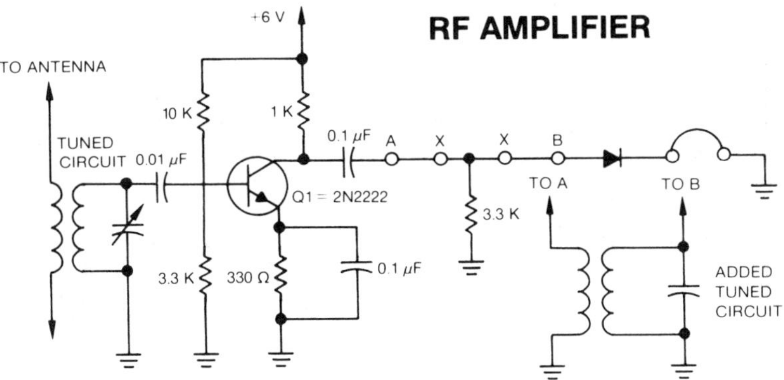

If you try this circuit, you will find that the sensitivity is much improved. You can make an even greater improvement by adding another tuned circuit at the detector to replace the 3.3-K resistor: break the circuit at points X and connect points A and B of the tuned circuit to points A and B of the circuit as shown. With two tuned circuits you will find the selectivity improved and that the detector works better.

Modern IC TRF Receiver

A modern single-TRF receiver can be constructed using a single IC to replace all of the RF amplifier, detector, and audio preamplifier functions previously performed by vacuum tube or discrete transistor circuits and numerous individual components.

The National LM172 is an eight-pin IC packaged in a round can (5/16 inch in diameter) and can be used for RF amplifier and detector functions at frequencies below 2 MHz. The circuit shown is used in the AM broadcast band (535–1,605 kHz) and has sufficient sensitivity for use in urban areas. The LM172 contains three wide-band amplifiers. One amplifier with the tuned circuit at its input is used as an RF amplifier. This tuned circuit also functions to replace the antenna (as will be discussed later). The first RF amplifier has a gain-control voltage input such that as the dc voltage increases on this input, the gain decreases. The output of the tuned-RF (TRF) amplifier is fed to a broadband (untuned) RF amplifier, as shown. The output of this broadband amplifier is then fed to the detector that demodulates the RF signal and also provides automatic-gain-control (AGC) voltage output. The detector output is fed to audio amplifier Q1, which drives the loudspeaker. The detected output is filtered by R4–C4 to provide automatic gain control for input signals greater than 50 MV at pin 2. As the carrier input tends to increase, the AGC voltage also tends to increase. This voltage, applied to the tuned-RF amplifier, tends to reduce its gain, which in turn tends to hold the carrier level constant.

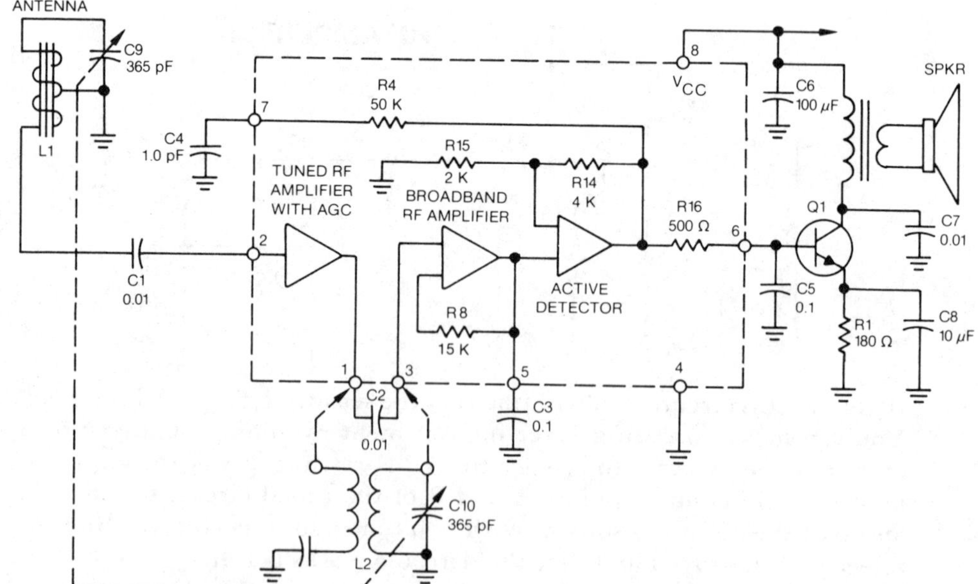

Modern IC TRF Receiver (continued)/The Limits of Receiver Sensitivity—Noise

With additional gain either preceding the module or between pins 1 and 3, it may be used for monitoring loran signals (1.8–2.0 MHz), or the numerous navigational and informational channels below 550 kHz. Since this particular circuit has only one tuned circuit (L1–C9), its selectivity is poor. The selectivity can be improved by adding a second tuned circuit between pins 1 and 3 of the IC, as shown in the previous figure by the dotted lines, instead of the coupling capacitor illustrated. The audio output at pin 6 is maintained between 2.1 and 2.4 volts by the AGC action and is used to bias the class-A audio amplifier Q1 directly for greatest circuit simplicity. Now let us move on to the next subject.

You may wonder why any number of amplifiers cannot be strung in cascade to provide any amount of gain desired—so that *any* signal would be detectable. The answer is that you can do this, but a point is soon reached beyond which no further improvement in sensitivity will occur. The reason for this is that it is not signal level *alone* that determines receiver sensitivity, but the *signal-to-noise ratio.* The signal-to-noise ratio (S/N) is the ratio of signal power to noise power at some point in the receiving system (typically, the output). Receiver noise comes from inside and outside the receiver. External noise comes from interfering signals, atmospheric effects (like lightning), operating electrical equipment, automobile ignition, etc. External interference or noise is sometimes collectively called *static.* Internal noise comes from irregularities—motion of electrons normally present in conductors or semiconductors—and is often called *thermal noise* or *hiss.*

SOURCES OF NOISE IN RECEIVERS

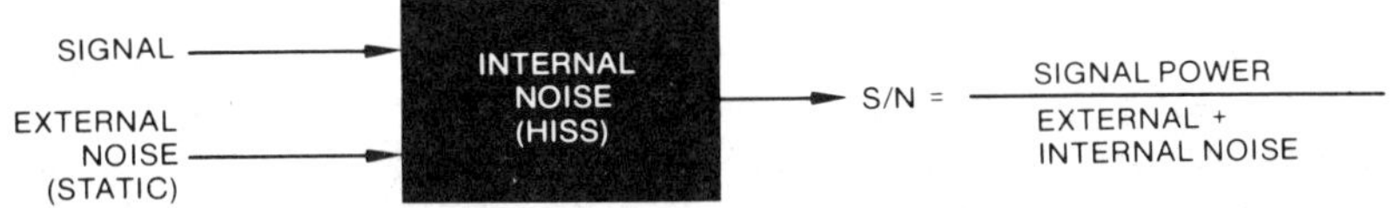

External noise predominates at frequencies below about 2 MHz, and becomes gradually less important, becoming negligible (except in extreme cases) above 30 MHz. Improvement in the receiver internal noise is of no value if it is below the external noise level. Most receivers currently in use below 2 MHz are limited by external noise almost entirely. You can hear this noise (static) on all AM broadcast receivers by tuning to a spot where no station is heard; the pops, hisses, and crackles of static are only too evident. At VHF and above, the internal receiver noise predominates, and hence keeping it at a minimum becomes important. This internal noise is due to the nonuniform flow of electrons in conductors and semiconductors and can be reduced—but never eliminated—by careful receiver design. You can hear this noise on an FM receiver tuned to where no station is heard. You'll hear a steady hiss due to internal noise.

The Limits of Receiver Sensitivity—Noise (continued)

Since both external and internal noise are *broadband*—that is, they are rather uniform over a broad range of frequencies—the signal-to-noise ratio can be improved by making the receiver as *selective* as possible, covering only the frequency bandwidth of the desired signal so that noise at other frequencies is excluded. For example, if a signal has a total bandwidth of 10 kHz, a receiver with a 20-kHz bandwidth will admit twice as much noise as is necessary and hence have a 2:1 poorer S/N ratio than a receiver with a bandwidth of 10 kHz. On the other hand, if the receiver bandwidth is narrowed to 2.5 kHz, there will be no improvement in the S/N ratio because signal is being excluded along with noise by about the same amount.

INTERNAL NOISE DEPENDS MAINLY ON THE FIRST STAGE NOISE CHARACTERISTICS

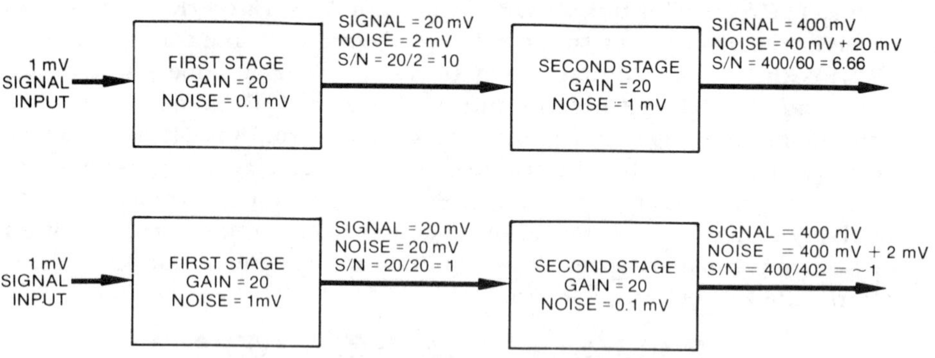

As you can see from the figure, it is the first-stage internal noise that is important in minimizing internal noise. Suppose you had two amplifier stages, each with a gain of 20. One produces the equivalent of 0.1 MV of noise at its input and the other produces 1 MV. As shown, if these stages are put in cascade to produce a gain of 400 (20 × 20), the S/N ratio at the output is radically different for the two configurations, depending on which of the stages comes first. Obviously, the first-stage noise level is most important. Since RF amplifiers can be made with a minimal noise contribution, their use is desirable to provide maximum sensitivity where thermal or hiss noise is predominant. Mixers in superheterodyne receivers tend to have much more internal noise than RF amplifiers and therefore RF amplifiers are more commonly used on receivers, particularly those for use above a few megahertz. The internal receiver noise is often specified for a given receiver in terms of the input signal necessary to give a specific signal-to-noise ratio at the output. For sophisticated receivers, the internal noise is specified in terms of *noise figure*, which is a measure of how far the receiver under test departs from an ideal receiver.

The Superheterodyne Receiver

The TRF receivers used in the early days of radio broadcasting were marvels of their time and far superior to the crude crystal sets. However, since all of the gain prior to demodulation was done at RF, and since early vacuum tubes had limited RF-gain capability, tube-type TRF receivers had limited sensitivity. Furthermore, the need to have multiple variable-frequency tuned circuits tracking precisely over a 3:1 (535- to 1,605-kHz) tuning range limited the degree of selectivity that could be obtained. The superheterodyne receiver, developed during World War I and introduced for broadcast use in the 1920s, satisfied the then vital need for a highly selective, sensitive receiver. Since the late 1920s, essentially all receivers are of the superheterodyne type. The superheterodyne receiver and FM were both invented by Major Armstrong, one of the truly great pioneers of early radio and radio broadcasting.

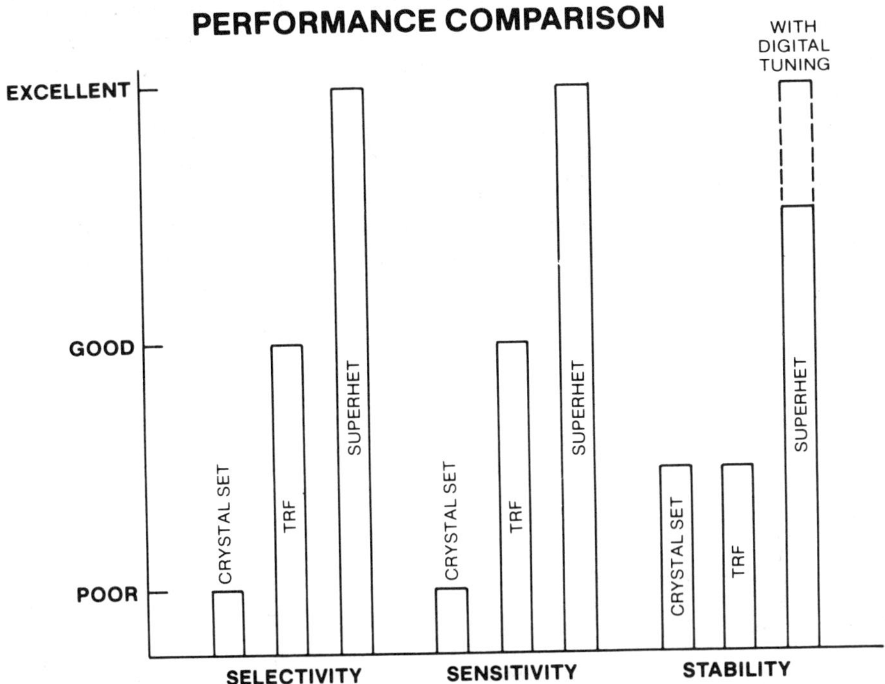

As you can see from the chart shown here, the superheterodyne receiver (often called a *superhet* for short) is greatly superior in all aspects of selectivity, sensitivity, and stability. The advantages of the superheterodyne receiver lie in the intermediate-frequency portion of the receiver. As you will learn, the RF amplifiers, detector, and audio circuits for a superhet receiver are no different than for a TRF receiver. The difference lies in the fact that in a superhet receiver, the incoming RF signal is converted to a fixed frequency (the IF), at which most of the amplification and selectivity are obtained. These can be carefully controlled because the IF amplifier needs to operate at only one frequency.

The Superheterodyne Receiver (continued)

As stated on the previous page, the superheterodyne receiver avoids the shortcomings of the TRF by converting the RF signal to a new, usually *lower frequency*, called the *intermediate frequency* (IF). This is accomplished by mixing the original RF signal with a second RF signal, generated in the receiver local oscillator (LO). The combination of the original RF signal with the LO signal produces the IF signal. As you remember from your study of transmitters in Volume 3, a mixer is a device that accepts two RF signal inputs (f1 and f2) and produces at its output two signals, the sum of the input frequency (f1 + f2) and the difference of the input frequency (f1 − f2). Usually the difference frequency is chosen as the intermediate frequency. This IF signal is identical to the original RF signal, except that it is displaced in frequency by the difference between the RF and LO signals (IF = RF − LO). The intermediate frequency is fixed because the difference between the signal input and LO frequencies is kept constant and does not vary, regardless of the frequency to which the receiver is tuned. This is done by having the LO frequency change with receiver tuning (ganged tuning capacitors) so that the constant difference is maintained regardless of the incoming signal frequency. Since the IF is fixed and lower than the lowest carrier frequency (455 kHz for standard AM broadcast), IF amplifiers can be designed to give rather large gains and high selectivity. The only parts of the superhet that differ from the TRF are the *local oscillator* and the *mixer* (the IF amplifier is actually a fixed-tuned lower-frequency RF amplifier).

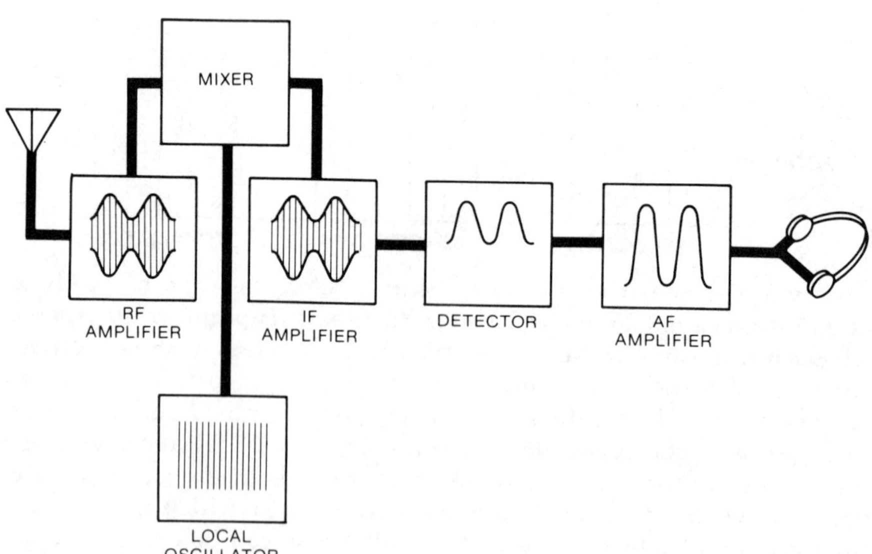

Selectivity of the Superhet—Signal Response

When you tune a superheterodyne receiver with an IF of 455 kHz to a station with a frequency of 880 kHz, you are setting the tuned RF circuit to 880 kHz and at the same time you are automatically tuning the local oscillator to 1,335 kHz. Two signals—one of 880 kHz, the other of 1,335 kHz—are fed into the mixer stage. The output of the mixer is one signal at a frequency of 455 kHz, which is the *difference* or *beat frequency* between its two inputs, and another of 2,215 kHz, the *sum* of 1,335 and 880. The 2,215-kHz signal is so far from 455 kHz that it is not amplified by the IF amplifier.

If, at the same time, the antenna picks up another station at a frequency of 1,100 kHz, that signal, if strong enough, could get by the first tuned circuit and would then be mixed with the local oscillator output in the mixer stage. This undesired signal of 1,100 kHz would produce a beat frequency of 1,335 − 1,100 or 235 kHz, which would not be amplified by the IF amplifier tuned to 455 kHz.

ACHIEVING SELECTIVITY

DESIRED SIGNAL	RF FREQUENCY (kHz)	OSCILLATOR FREQUENCY (kHz)	MIXER OUTPUT FREQUENCY (kHz)	FREQUENCY SELECTED BY IF AMPLIFIER
→	880 1,100	1,335 1,335	455 235	←
→	880 1,100	1,555 1,555	675 455	←

The IF amplifier tuning does not vary. It is always tuned to 455 kHz, so you can see that only the *difference* signal produced by the desired station (880 kHz) will be amplified by the IF amplifier. Since the *sum* signal produced by an undesired signal of 1,100 kHz produces a beat signal whose frequency is different from the intermediate frequency, that beat signal is not amplified. Thus, the superhet has selected the proper input signal on the basis of the frequency of the difference signal produced in the mixer stage.

In order to hear the 1,100-kHz station, the receiver would have to be *retuned*. Turning the knob changes the frequency to which the RF amplifier is tuned and, at the same time, changes the local oscillator frequency. A two- or three-section ganged tuning capacitor does the trick. Tuning the receiver does not affect the IF stages. When the RF and mixer tuned circuits are set at 1,100 kHz, the ganged capacitor section that tunes the oscillator is at the right position to put out a signal of 1,555 kHz; the intermediate frequency remains at 455 kHz.

Now it is the 1,100-kHz input signal that produces the 455-kHz beat frequency. The beat produced by the 880-kHz signal would be the difference between its frequency and the 1,555-kHz local oscillator frequency, or 675 kHz, which will not be amplified by the IF stages.

Selectivity of the Superhet—Image Response

In order for the superhet to work properly, the local oscillator must be adjusted so that it will always tune to a frequency at a fixed difference from the desired RF signal. Thus, as the receiver—that is, the RF and mixer tuned circuits—are tuned from 550 to 1,600 kHz, the local oscillator should tune from 1,005 to 2,055 kHz. Then, any signal picked up at the frequency to which the receiver is tuned will produce an IF signal of 455 kHz (which is the standard intermediate frequency for AM broadcast receivers). Since all of the IF gain is implemented at a common fixed frequency, all of the stages can be made a fixed narrow bandwidth, and a high degree of selectivity is thereby obtained.

Perhaps you have already noted that there is a problem with the selectivity of the superhet receiver. On the previous page it was stated that the intermediate frequency is the difference between the signal frequency and the local oscillator (LO) frequency. For example, for a station at 880 kHz the local oscillator is set at 1,335 kHz to produce an IF of 455 kHz. Note, however, that an input signal at 1,790 kHz will produce a difference between the local oscillator frequency and the signal exactly equal to the IF (1,790 − 1,335 = 455 kHz). This undesired signal is called the *image*, and it is always located at a point twice the value of the IF away from the desired signal.

SELECTIVITY OF THE SUPERHET

DESIRED SIGNAL	IF	LOCAL OSC
880 kHz	455 kHz	1,335 kHz
90.3 MHz	10.7 MHz	101.0 MHz

IMAGE
1,790 kHz
111.7 MHz

Obviously if the LO is on the low side, the image is below the desired signal by twice the IF. The only discriminants against the image are in the RF circuits. For example, the mixer input tuning will provide some selectivity against the image. At lower signal frequencies, the selectivity of the mixer input circuit is adequate to keep the image responses down; at signal frequencies above the broadcast band, however, the additional selectivity of a tuned RF amplifier is important in reducing the image response. You will also notice later that as higher radio frequencies are used, the intermediate frequency chosen for the IF amplifier is increased to move the image further away from the desired response. For example, a broadcast receiver has an IF of 455 MHz, while a CB radio (at 29 MHz) may have an IF of 4.5 MHz, and an FM receiver (at 100 MHz) may have an IF of 10.7 MHz.

Selectivity of the Superhet—Double Conversion

Since the selectivity and gain of an IF amplifier depend on the frequency (both better at lower IFs)—a low IF is desirable for some applications. On the other hand, if the received signal is broadband, or if the input frequency is higher, a higher IF is desirable because broadband circuits are easier to design at these higher frequencies. For example, TV receivers (to be discussed later) require an IF bandwidth of almost 6 MHz; for these, the IF is between 40 and 50 MHz. Radar receivers commonly use 30 MHz, 60 MHz, or even higher IFs for the same reason. When a combination of high frequency (VHF and above) and high selectivity is desired, *double* conversion is often used. For such applications, there are two IF amplifiers—one with a high IF to provide best image rejection and a second one with a low IF to provide best selectivity.

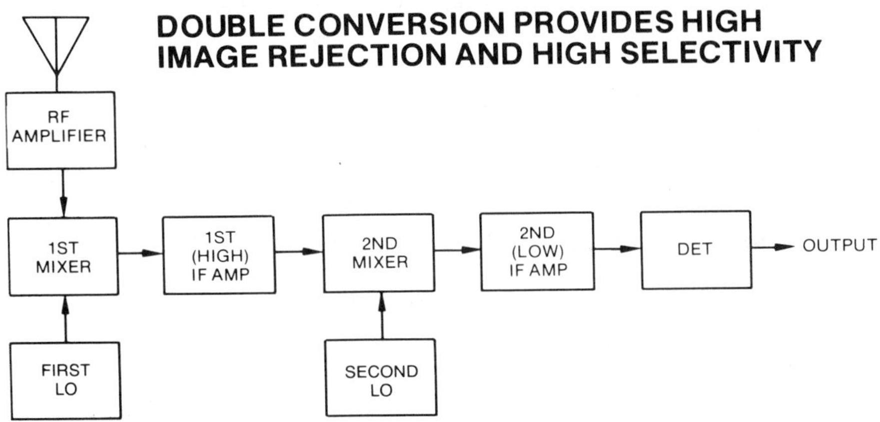

The double conversion superhet has two mixers and local oscillators. The first mixer takes the high-frequency received signal and mixes it with the signal of a high-frequency (first) LO to produce a high-frequency (first) IF (to minimize image response). This is amplified (or in some cases just bandpass filtered) and applied to the second mixer, which mixes it with a lower-frequency (second) LO to produce a second (low-frequency) IF (for selectivity). In communication receivers, when tuning range is narrow, the first LO is crystal controlled for better stability at the higher frequency. The circuits between the first and second mixers must then be broadband, because the first IF must vary over the tuning range (because the first LO is fixed). The second LO is then made variable and used for tuning; the lower frequency allows greater stability for this. Tuning ranges over different frequencies are obtained by providing several selectable first-LO crystal frequencies.

When the tuning range is too great for this, the tuning must be done in the first LO, with the second LO fixed and usually crystal controlled. As you will learn, special crystal filters are also used in communication receivers to improve their selectivity.

Modern IC Superhet Receiver

The above can be constructed using a single IC to perform the oscillator, mixer, IF amplifier, detector, and audio preamplifier functions.

The RCA CA3088E is an IC suitable for AM receiver applications up to 30 MHz. Since many high-frequency receivers use IF amplification at the higher frequencies, the high-frequency characteristics can be very important. In the AM broadcast receiver shown, double-tuned transformer-coupled circuits are used in the IF. However, other forms of bandpass filters (such as ceramic filters), which you will learn about later, would be equally suitable. Because of the size (the CA3088E is a single IC chip), the IF and mixer tuned circuits are external to the chip, as are some of the larger audio-coupling components and the audio volume control. The CA3088E contains both RF and IF AGC functions, and a tuning-meter driver that furnishes a buffered AGC output voltage to use with a tuning meter. A dc voltage, internally regulated by a zener diode, supplies the RF amplifier, converter, and first IF amplifier. You can see how simple an IC of the proper design can make something as complicated as a complete superhet receiver very simple from an external-circuit standpoint.

TYPICAL AM BROADCAST RECEIVER USING THE CA3088E

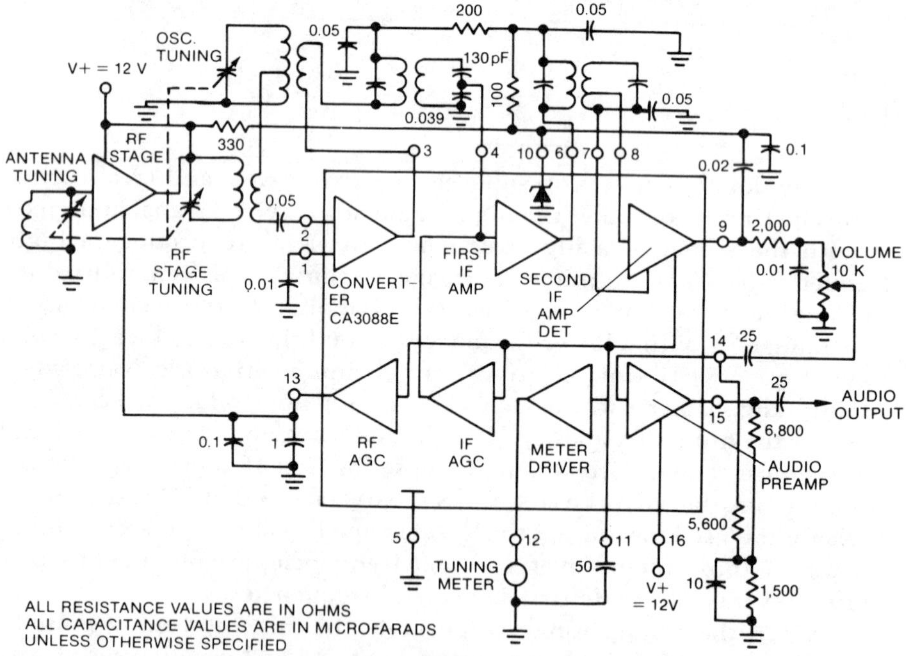

The IF amplifier output drives the detector, which strips off the modulation and also can provide for AGC. The detector output is fed to the external volume control, and the volume control output is fed to an audio preamplifier (included in the IC). The output of the audio preamplifier is delivered from the IC for input to an audio power amplifier.

Review of Radio Reception and Receiver Characteristics

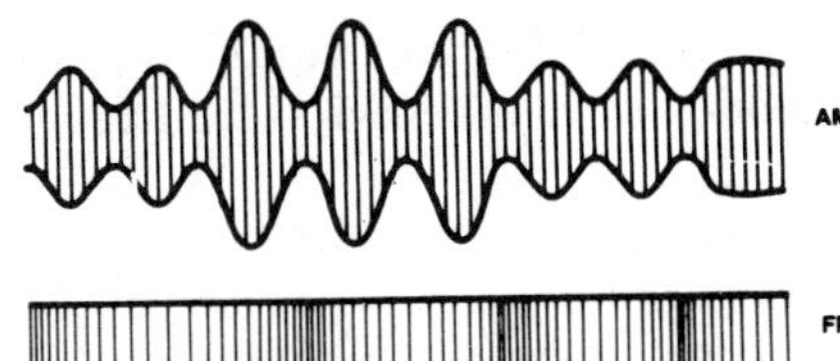

1. THE TWO BASIC MODES OF RECEPTION are amplitude modulation (AM) and frequency modulation (FM).

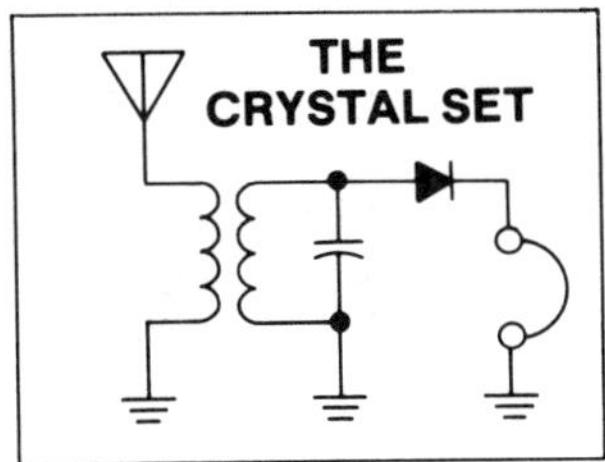

2. A DIODE DETECTOR rectifies the received signal to develop a pulsating dc signal output whose envelope is the waveform of the desired intelligence.

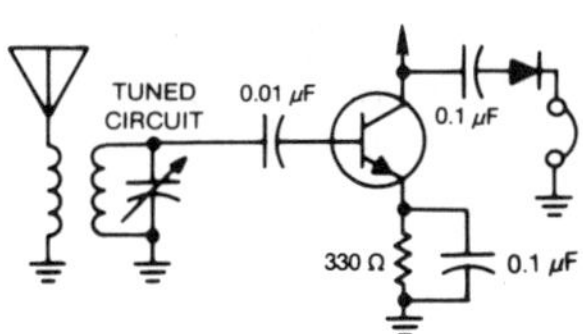

3. THE SENSITIVITY AND OUTPUT of a detector can be increased by preceding it by one or more RF amplifier stages, forming a tuned radio frequency (TRF) receiver.

SOURCES OF NOISE IN RECEIVERS

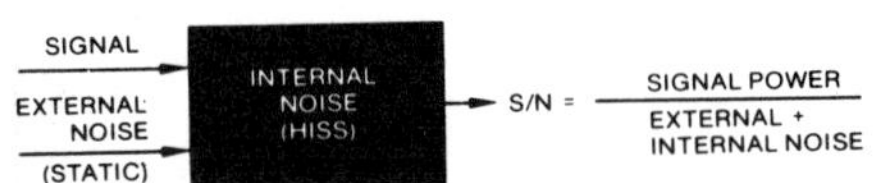

4. NOISE generated in its circuits affects a receiver's ability to receive weak signals—especially in the circuits in the front end.

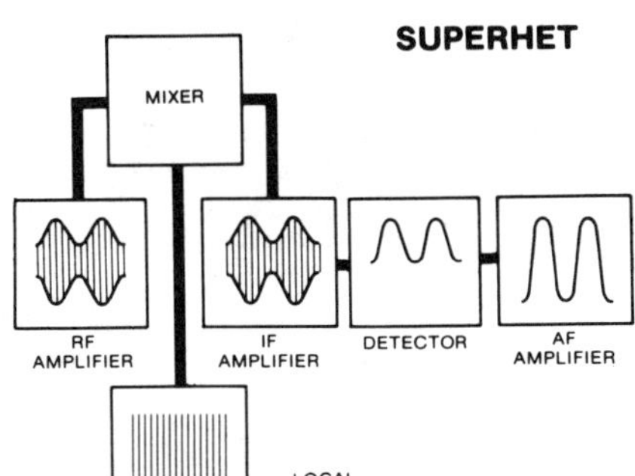

5. IN A SUPERHETERODYNE the received signal is mixed with a local oscillator signal to produce an IF, at which optimum amplification and selectivity are possible.

Self-Test—Review Questions

1. Briefly list the elements of a receiving system and describe their function.
2. List and identify the FCC-designated modes of transmission and reception.
3. Describe the operation of a simple diode detector.
4. How does a crystal set work?
5. Draw a block diagram of a TRF receiver. Describe its operation.
6. Draw a block diagram of a superhet receiver. Describe its operation.
7. Compare the TRF and superhet receivers. What are their advantages and disadvantages?
8. What causes an image in superhet receivers? How do RF selectivity and double conversion minimize image responses?
9. What factors limit receiver sensitivity? How can it be maximized?
10. Compare the role of the second detector in a superhet to the detector in a TRF receiver.

Learning Objectives—Next Section

Overview—In the next section you will study various types of antennas and then proceed to a detailed discussion of AM radio receivers, including their contents and characteristics.

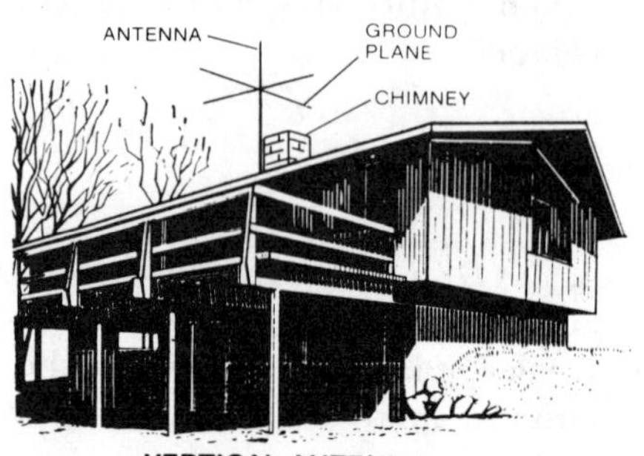

VERTICAL ANTENNA

HORIZONTAL ANTENNA

Receiving Antenna Functions

Just as transmitting is the mirror image of receiving, so receiving antennas are the mirror of transmitting antennas.

The purpose of the receiving antenna is to intercept the electromagnetic waves radiated from the transmitting antennas. When these waves cut across the receiving antenna, they generate a small voltage in it. This voltage causes a weak current to flow in the antenna-ground system. This feeble current has the same frequency as the current in the transmitting antenna. If the original current in the transmitting antenna is modulated, the receiving antenna current will vary in exactly the same manner. This weak antenna current, flowing through the antenna coil, induces a corresponding RF signal in the input circuit of the receiver.

RECEIVER ANTENNAS INTERCEPT THE RADIO WAVES SENT OUT BY THE TRANSMITTER

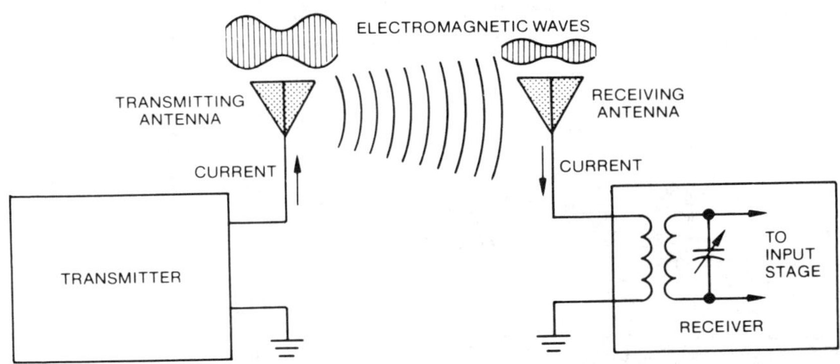

A receiving antenna should feed as much signal and as small an amount of undesired interference as possible to the receiver. It should be constructed so that the signal is not lost or dissipated before reaching the receiver. It should give maximum response for the frequency or band of frequencies to which the receiver is tuned. Thus, the condition of match and other parameters specified for transmitting antennas also apply for receiving antennas; however, the requirements are usually much less critical because receivers can be made comparatively sensitive and do not handle significant power. An antenna can also be *directional*, which means that it will give best response in the direction from which the operator wishes to receive.

The receiving-antenna problem is more easily solved when the receiver is operated in conjunction with a transmitter or with a transceiver, since the transmitting antenna is usually designed to incorporate the desirable features that have just been listed, and the *same* antenna is used for both transmitter and receiver. A switch or relay is used to connect the antenna to either the transmitter or receiver, depending on the operating function at a particular moment. However, when no transmitter antenna is available, it is necessary to use a *separate* receiving antenna, paying attention to the four considerations of noise, signal loss, operating-frequency response, and directivity.

Receiving Antenna Types

The simplest receiving antenna is merely a piece of wire of arbitrary length. In a field of a given strength (volts/meter), the longer the wire, the greater the voltage induced. Thus, one would be tempted to build an antenna of the greatest possible length. On the other hand, such an antenna could be very difficult to match so that the losses could outweigh the gains. A *resonant* antenna is desirable because it looks resistive and hence is easiest to match, particularly over a band of frequencies. A resonant antenna is one that has a length that is a multiple of a half wavelength. The basic antenna is the *half-wave dipole*—that is, a wire that is a half wavelength long. In theory, a half wavelength (in feet) is about equal to 500/f where f the frequency, is expressed in megahertz. As a practical matter, end effects and other phenomena cause the antenna to appear longer than it actually is physically, so a factor of 470/f is more realistic at frequencies below about 50 MHz. A single-wire half-wave antenna looks, at its center point, like a resistive impedance of about 70 ohms and is easy to match.

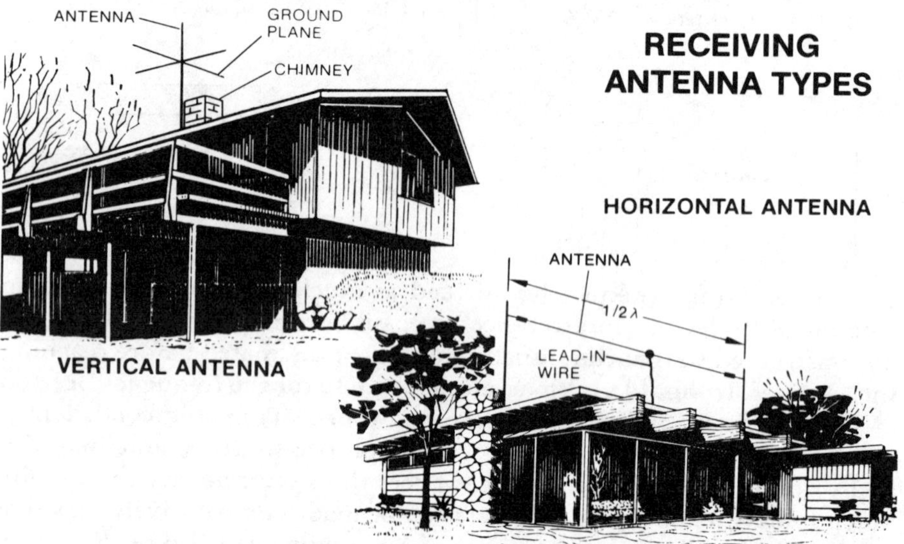

Usually, a half-wave antenna is *horizontally polarized*—that is, the main axis of the antenna is horizontal. For some applications, as you will see, *vertically polarized (vertical)* antennas are used, as you will recall from your study of transmitting antennas. In any case, for a resonant antenna the voltage is minimum at the center and maximum at the ends, and the current is maximum at the center and minimum at the ends. If the antenna is fed at the center it is called a *current-fed antenna*. If it is fed at the end, it is called *voltage-fed*.

Usually, horizontal receiving antennas are used for short-wave applications and for medium-wave applications, as well as for TV and FM broadcasting. Vertical antennas are used for receiving in portable and in communications equipment.

Receiving Antenna Types—Broadcast/Short-Wave Receivers

The wavelength (λ) used for AM broadcast ranges from 200 to 600 meters. Antennas in the order of $\lambda/2$ are thus not practical. For home radio use, the configuration is almost always some variation of a loop antenna, as shown in the diagram. Loops operate somewhat differently from conventional wire antennas. Since the loop circumference is small compared to the wavelength, uniform in-phase currents exist around the loop conductors and the reception properties are relatively independent of loop geometry, and the antenna pattern for a small loop is constant in the plane of the loop. In the plane perpendicular to the loop, the voltage at the antenna terminals varies from a maximum off the sides of the loop to a sharp minimum or "null" in the direction perpendicular to the loop plane. The pattern is shown in (D). This directional property is quite noticeable when receiving weak stations; with strong stations, however, receiver AGC action compensates for the effect of θ variation in received signal amplitude, except in the nulls.

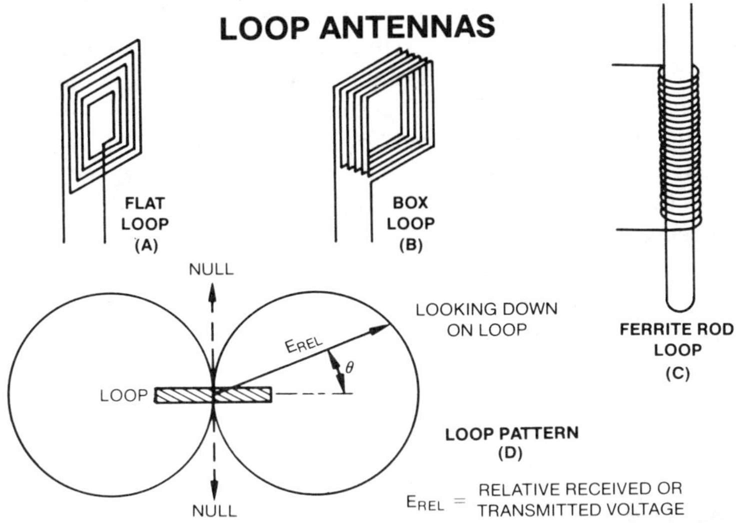

The ferrite *loopstick* antenna (C) is used in most AM radios and consists of a ferrite rod or slab wound with many turns of wire. The ferrite core intercepts the radio-wave magnetic field to generate an electromagnetic field in the core, and the variations of this core field, at the radio-frequency rate, convert the field to a variable current in the antenna coil surrounding it. This produces an RF voltage at the input of the receiver first stage. Also, the antenna circuit is usually tuned.

Because of the directional properties of loop antennas, they are used in *direction-finding equipment* to locate the direction from which a broadcast is coming. This is done by rotating the loop until the signal is *weakest* (or at null), and the direction then is perpendicular to the plane of the loop. The reason that the null is used is because it is much sharper and easier to find than the peak. Direction-finding receivers have the antenna mounted on a compass scale, so that direction can be read directly.

Receiving Antenna Types—Mobile Systems

Mobile operation in automobiles, trucks, and boats have special problems for operation at wavelengths at 30 MHz and below because of the physical length of the antenna required to achieve a half wavelength. Since an automobile is shielded on the inside, an external antenna is always required. Usually, mobile antennas take the form of a vertical rod or *whip* with a length much less than half a wavelength for frequencies below 30 MHz. Generally, antennas for mobile operation are very short for operation in the AM broadcast band, but approach a quarter wavelength at 30 MHz (Citizens Band). As you learned in your study of transmitting antennas, an antenna less than a half wavelength long looks capacitive (as well as resistive). For quarter-wave antennas with a ground plane, the same is true. To match a short antenna, an inductive reactance from the source is required. Alternatively, the antenna can be electrically lengthened by adding inductance in series with its output. Since these antennas are fed at a high-current point, a low-resistance (loss) coil is important if efficiency is to be maximized.

At VHF and UHF, it is possible to use antennas that are a quarter wavelength long with a ground plane that gives the equivalent of a half-wave antenna. Thus, VHF-UHF antennas do not generally need reactance in the line to achieve a matched system. As you remember from your study of antennas in Volume 3, the ground plane acts as a mirror that allows for a quarter-wave antenna to function as a half-wave antenna. To be effective, the ground plane must extend more than a quarter wavelength. On automobiles, the roof can act as a ground plane for some VHF and UHF applications. You often see CB antennas mounted on chimneys or towers with ground plane radials, as shown in the figure.

MOBILE ANTENNAS

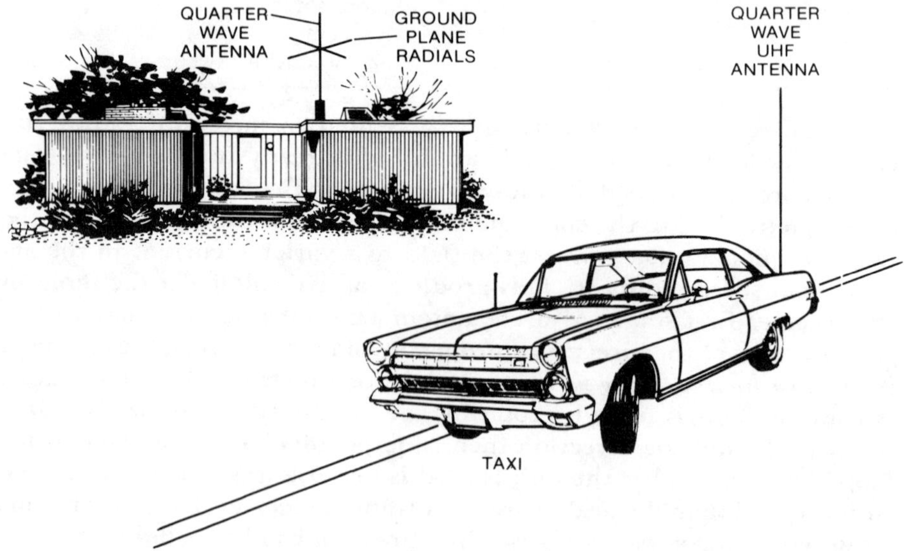

Receiving Antenna Types—FM and TV

FCC regulations in the United States require FM and TV stations to radiate horizontal polarization, but circular or elliptical polarization may be employed if desired. Most FM and TV stations transmit both horizontal and vertical polarizations, since this allows good reception with either horizontal rooftop antennas or with vertical whips and loop antennas. Communication stations usually use vertically polarized whip antennas for both transmission and reception.

Since VHF and UHF transmission is limited to line-of-sight, superior performance, especially in fringe areas, can be obtained with an elevated or rooftop antenna. Antennas can be either omnidirectional, for receiving signals from many directions, or directional with high gain, for receiving signals primarily from one direction.

Omnidirectional horizontally polarized VHF-UHF antennas usually take the form of a pair of horizontal crossed-folded dipoles. As you learned in your study of transmitting antennas, each individual dipole has a figure-8 pattern, and the sum of the crossed pair has nearly uniform gain in the horizontal direction. A single folded dipole bent into an S-shape can also provide fair approximation of omnidirectional coverage.

FM/TV RECEIVING ANTENNAS

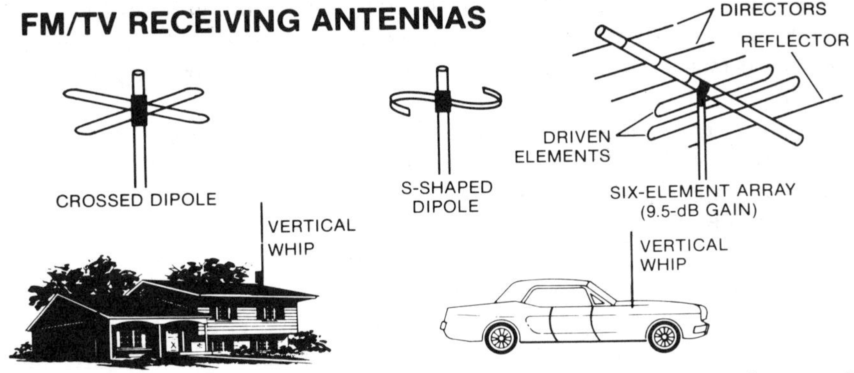

A directional receiving antenna provides for improved reception in one direction at the expense of other directions. If all (or most) of the transmitting stations are located in the same direction—the direction of the city or urban area—then the high gain associated with high antenna directivity can make the difference between good or poor reception. Most directional antennas are used for TV reception and consist of an array of dipoles, with one or more folded dipoles being used as the driven elements, and additional directors and reflectors added to increase the gain and directivity. These additional elements are referred to as *parasitic* elements, and they are coupled to the *driver* (folded dipole) antenna by the mutual space impedance between them. The current induced in the parasitic antenna will produce a radiation field which combines with that of the driver elements to yield the overall pattern. By choosing the number, spacing, and length of the parasitic elements, it is possible to control the gain, directivity, and bandwidth of the array. These directional arrays can be used horizontally or vertically.

RF Amplifier Stage

In a superhet receiver, the signal from the antenna may be connected directly to the mixer stage input, without using an RF amplifier stage. Except at low frequencies, however, you will find receivers usually contain stages of RF amplification preceding the mixer. The RF amplifier in these cases is performing three functions that improve receiver performance.

The major function of the RF amplifier is to *improve the signal-to-noise ratio.* The mixer stage usually produces much more internal noise than an RF stage of amplification. The signal, plus the mixer noise, is amplified by the following IF amplifier stage. As you know, a low-noise amplifier preceding a noisy amplifier provides a better overall signal-to-noise ratio. Thus, an RF amplifier will improve the overall signal-to-noise ratio in a superhet receiver. In some cases, two RF amplifier stages are used; usually, however, a single stage is sufficient to provide for good performance.

The second function of the RF amplifier stage is concerned with *selectivity.* You will recall that in the TRF receiver, the RF amplifier stages enabled the operator to select the desired signal from a group of signals whose frequencies were very close to each other. The RF amplifier in a superhet serves to prevent interference from the image signal whose frequency is twice the value of the IF away from the desired signal. Since the basic selectivity to signals at nearby frequencies is obtained in the IF amplifier, the selectivity of the RF amplifier is most important in terms of how well it rejects the image. In addition, the RF amplifier selectivity helps in keeping strong local signals at nearby frequencies from overloading the receiver.

The third function of the RF stage is related to *radiation from the oscillator.* It should not be forgotten that this oscillator is a low-powered transmitter. If there is no RF amplifier stage, the oscillator is connected through the mixer stage to the antenna. This antenna can radiate some energy from the oscillator and cause interference with reception in other nearby receivers at the local-oscillator frequency. This radiation may be reduced or prevented by using one or more stages of RF amplification, and by carefully shielding the oscillator stage.

BASIC FUNCTIONS OF RF AMPLIFIERS

- IMPROVES SIGNAL-TO-NOISE RATIO
- IMPROVES SELECTIVITY AND REJECTS IMAGE
- REDUCES LOCAL OSCILLATOR RADIATION

Local Oscillator

As you have already learned, the function of the local oscillator (LO) in a superhet receiver is to provide an unmodulated reference signal that *tracks* the desired RF signal at a fixed, offset frequency—typically, 455 kHz for the AM band. Local oscillators have circuits that are the same as the oscillators you studied previously, and these oscillators are very low in power output. Below are examples of typical operating frequencies.

f_o FREQUENCY OF RF CARRIER	f1 LOCAL OSCILLATOR FREQUENCY	INTERMEDIATE FREQUENCY
550 kHz	1,005 OR 95 kHz	455 kHz
1,000 kHz	1,455 OR 545 kHz	455 kHz
1,600 kHz	2,055 OR 1,145 kHz	455 kHz

The LO used in AM broadcast receivers and other low- or medium-frequency receivers is usually 455 kHz above the RF carrier, or 995 to 2,055 kHz for the case shown. Thus, while the RF frequency is changing by a factor of about 3:1 (540–1,600 kHz), the LO must change by about 2:1 (995–2,055 kHz) to maintain perfect tracking. This is done by controlling the relative size and shape of the plates of the two sections of the ganged tuning capacitor where capacitive tuning is used, and by relative tuning slug motion where variable-inductance tuning is used.

As you learned previously, tracking between tuned circuits can be maintained with appropriate care and the addition of trimming capacitors, etc. Similarly, the ganged-tuning capacitor used in a superhet receiver can be designed to accomplish the same objective. Usually, the LO capacitor is smaller and designed to tune an inductor over the desired LO range, while the RF tuning capacitors tune simultaneously over the desired RF signal input range.

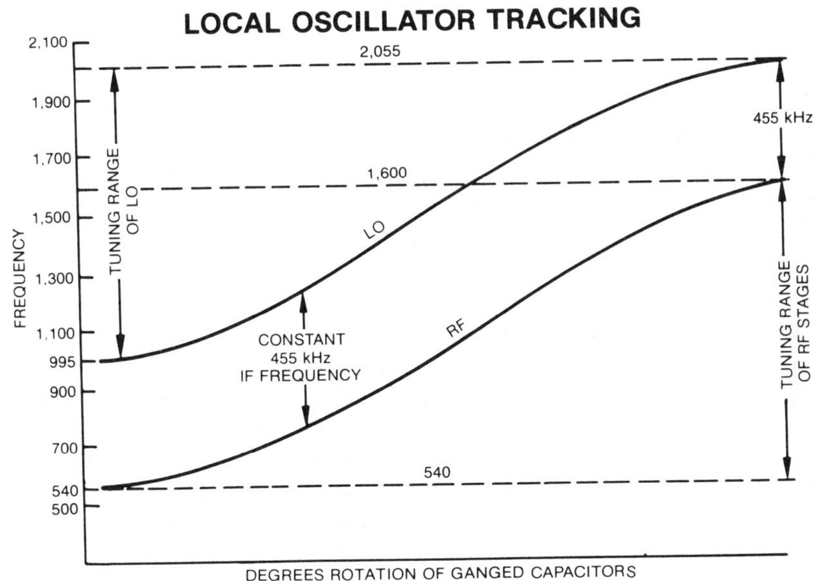

The Mixer—What a Mixer Does

The mixer is used in a superhet receiver to convert the incoming RF signal to the intermediate frequency. It must perform this frequency translation without distorting the incoming signal. As you know from your earlier study of mixers in Volume 3, the mixer accepts two input frequencies—the received signal and the local oscillator signal. The mixer *output* contains four different frequencies: (1) the RF signal, (2) the LO signal, (3) the sum of 1 and 2, and (4) the difference between 1 and 2. The difference signal is usually the desired signal in receiver mixer application. If the mixer is linear—that is, the output difference signal is directly proportional to the incoming RF signal—then the difference signal has all the information contained in it. Tuned circuits allow the desired signal (4) while discriminating against 1, 2, and 3 above.

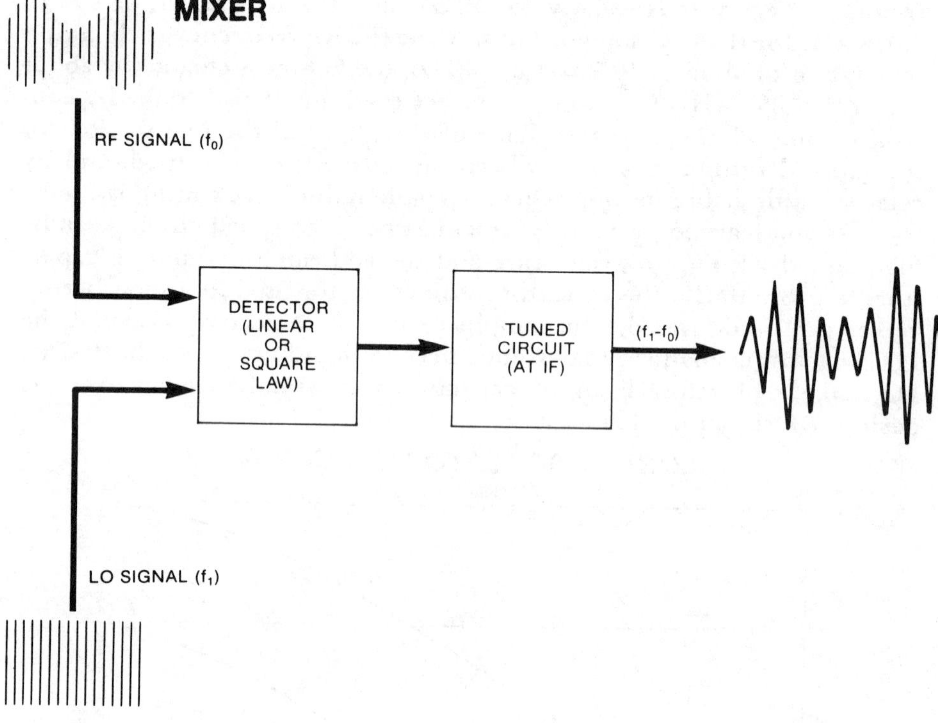

The desired IF signal is filtered by the primary winding of the first IF transformer included in the collector circuit of the mixer. The secondary winding of this IF transformer is coupled to the base of the first IF amplifier stage.

Mixers are sometimes called *frequency converters*, *frequency translators*, and *heterodyne detectors* or *first detectors*. The mixing or *beating* of two signals together in a mixer is sometimes called *heterodyning*, and this is the origin of the word *superheterodyne*.

Experiment/Application—Mixer Operation

You can see how mixers work by using a very simple test setup. You will need two signal generators, with at least one capable of tone modulation. These will be used as the signal and LO inputs. Suppose you connected the circuit shown below:

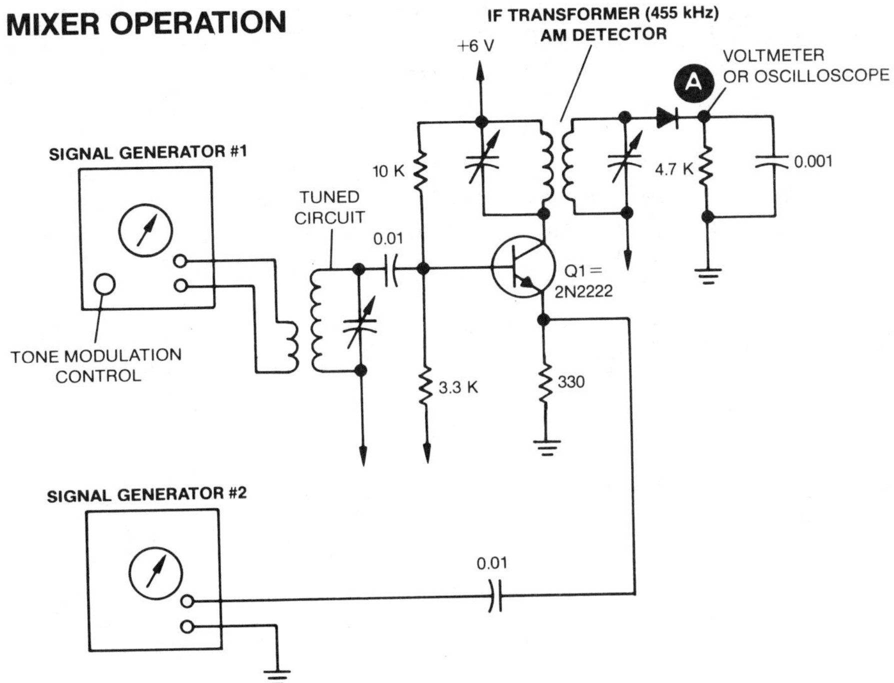

Set the output from signal generator 2 to maximum output. Set the output from signal generator 1 at least 10 times lower. Tune the system up to 1 MHz by setting signal generator 1 to 1,000 kHz and signal generator 2 to 1,456 kHz. Tune up the tuned circuit for maximum response at the detector output (maximum dc level). Then tune the IF transformer for maximum response at 455 kHz. If you tone modulate signal generator 1, you will find an ac voltage developed at the detector output. You can hear the tone if you replace the RC at the output of the detector by a pair of headphones or connect an oscilloscope at point A, as shown.

Without changing the signal generator outputs, tune signal generator 1 to 544 kHz. As you know, this is the image frequency. Compare the response at the correct and image frequencies. This will give you an idea of how much image rejection you get from the single tuned circuit at the input.

You can check the linearity by plotting the change in detected output level as the percentage tone modulation is varied. As you will see, the signal produced at the IF is a faithful (linear) reproduction of the tone-modulating signal.

IF Amplifier Operation

The intermediate-frequency amplifier is fixed tuned to the constant difference in frequency between the incoming RF signal and the local oscillator. As you know, this difference signal comes from the mixer. The tuning of the IF amplifier stage is usually accomplished by means of tuned IF transformers. The one associated with the amplifier input circuit is called the *input* IF transformer, while the one associated with the amplifier output circuit is called the *output* IF transformer. If more than one IF stage is used, *interstage* IF transformers are also employed between IF stages.

Since this amplifier is designed to operate at only one fixed frequency, the IF circuits may be adjusted for high selectivity and maximum amplification. It is in the IF stage that *practically all* of the *selectivity* and *voltage amplification* of the superhet are developed.

THREE-STAGE IF AMPLIFIER TUNED CIRCUITS PROVIDE SELECTIVITY AND AMPLIFIERS SUPPLY SENSITIVITY

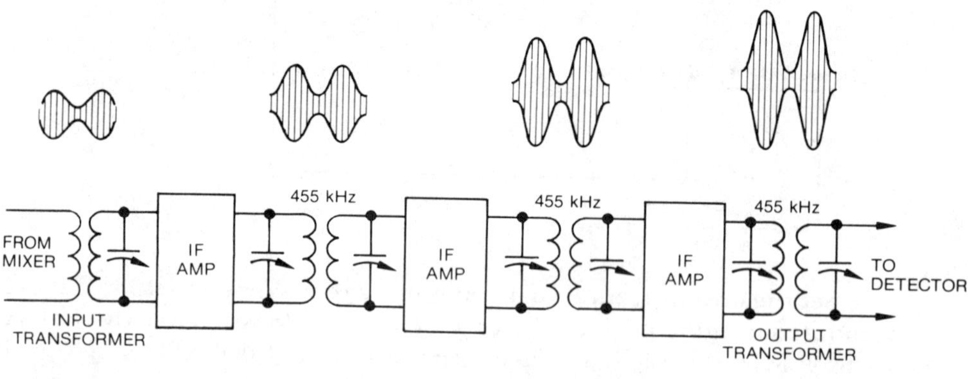

The intermediate frequency used in most AM broadcast superhet receivers is 455 kHz, although other intermediate frequencies are also used, as described previously. Using a low intermediate frequency, such as 175 kHz, results in high selectivity and voltage gain, but also increases the possibility of image-frequency interference. A high intermediate frequency reduces the possibility of image interference, but requires higher Q IF transformers and more stages to achieve the desired selectivity. In many cases, you will find the IF transformers tapped. This is done to minimize loading of the tuned circuit by the low transistor input impedance that can reduce the Q and hence the selectivity of the tuned circuit. The output of the last IF stage is fed to a detector that you studied earlier to recover the modulating signal.

In many high-performance superhet receivers used for communications applications and FM reception, crystal filters are used to provide very good control of the IF passband response. These will be discussed later in this volume.

Automatic Gain Control

As you know, the detector output generates a dc signal whose amplitude is proportional to the carrier signal level. This dc signal can be used to provide *AGC*, or *automatic gain control.*

AGC is usually applied to the first IF amplifier in a superhet receiver, and its function is twofold. First, AGC tends to keep the *receiver output constant* as the signal fades—due to either atmospheric conditions or changing tuning to a weaker station. And second, it *prevents amplifier overload* on a strong signal. If an RF amplifier is used, the AGC is also applied to the RF stage. When RF AGC is employed, it is usually in the form of *delayed AGC*. Delayed AGC is used to avoid reduction of RF amplifier gain on weak signals, which can cause a loss of sensitivity because reduced gain can degrade the signal-to-noise ratio, as you learned previously. Delay is accomplished by requiring the detector output to rise above a fixed threshold level before the RF amplifier gain is reduced.

The block diagram shows a typical AGC system with delayed AGC for the RF amplifier. Typically, the AGC is applied to the base or emitter of the transistor to reduce the quiescent current through the transistor and hence the gain. As you will learn, this approach to gain control is not practical for broadband amplifiers, because bias changes cause large changes in transistor input capacitance and hence can cause detuning.

The RCA CA3088 IC provides both IF and delayed RF AGC. In the performance curve, the IF gain is continuously reduced as signal level increases, but the RF AGC voltage is virtually constant for IF input levels of less than 2 mV (delayed) and then increases rapidly for high-level signals.

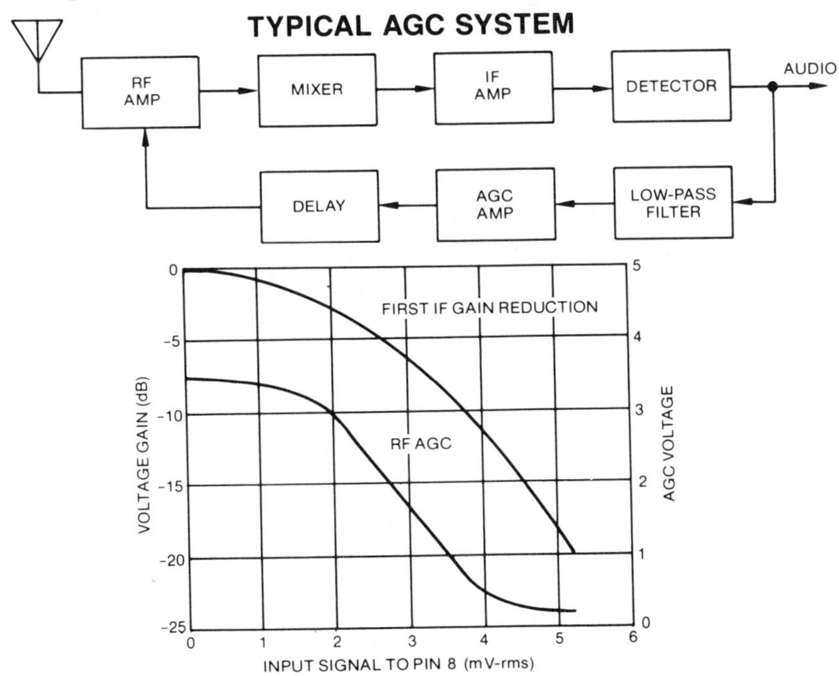

Automatic Gain Control (continued)

The AGC function can also be implemented by using a transistor as a variable resistance—as it is in the LM172 IC you studied when learning about the TRF receiver. This approach avoids the detuning effects of simple changes in transistor bias. In the LM172, an emitter-coupled transistor pair (Q2–Q3) is used as a series-shunt attenuator. The base of Q1 is held at +2.1 volts (dc) by the drop across diodes D1 through D3, and the base of Q2 is held at 1.4 volts (2.1 volts $- V_{BE2}$). Since the emitter of Q2 (and Q3) is held at +0.7 volt, the AGC voltage must be equal to or greater than 2.1 volts ($V_{BE3} + V_{BE4}$ + 0.7 volt) in order for Q3 to conduct. For AGC voltages less than 2.1 volts, Q3 is cut off and Q2 functions as a simple emitter follower. When V_{AGC} equals 2.1 volts (3 V_{BE}), Q2 and Q3 conduct equally and share the current through *source* resistor R2. As the AGC voltage tries to increase beyond 3 V_{BE}, Q3 turns on harder—conducts more—and Q2 starts to turn off—conducts less. Thus, the effective emitter resistance of Q2 (R_{E2}) increases, and the emitter resistance of Q3 (R_{E3}) decreases as the AGC voltage tries to rise above 2.1 volts. The sketch considers the emitter-coupled Q2–Q3 as a shunt attenuator adjusted by the signal to keep RF output constant.

AGC SYSTEM

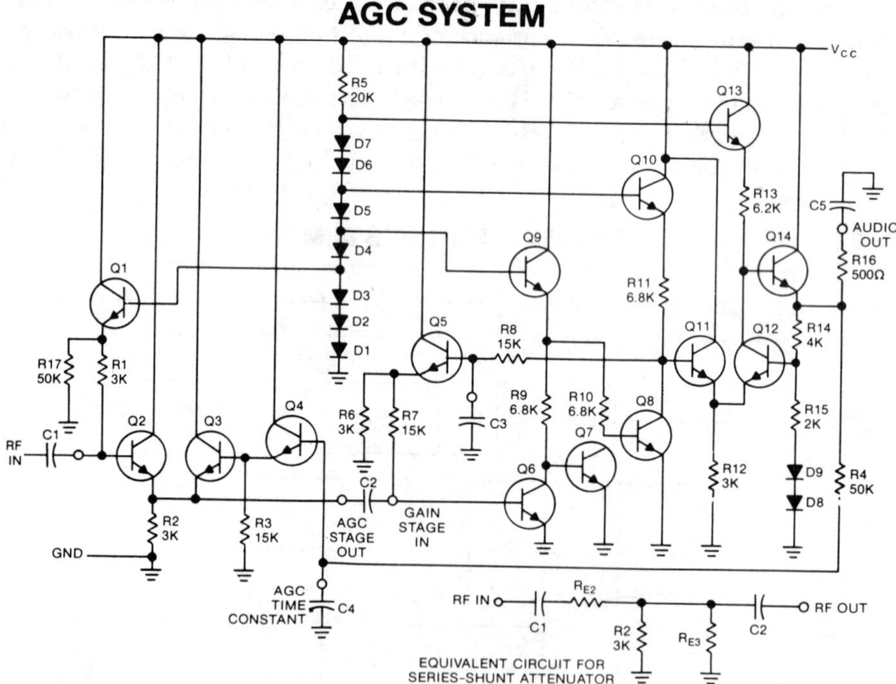

In any AGC system, care is necessary to keep the AGC free of any audio-output components. If the AGC filter allows low-frequency audio signals through, the receiver gain will be varied at that audio rate and the modulation will be erased, resulting in loss of low-frequency response. On the other hand, if the AGC frequency response is too slow, it will not respond rapidly enough to fading or station changes.

Experiment/Application—AGC Operation

Essentially all AM broadcast receivers use AGC. It would be somewhat difficult to build such a receiver from parts, but it could be done. However, AM broadcast receivers are very common, and you can use a commercial AM receiver to advantage to study AGC action. To do this, you must be able to get at the circuits. Assuming that you can do so, obtain an AM receiver and locate the AGC line. *Do NOT use an ac/dc radio for this experiment.* You can find it most easily by locating the second detector and tracing the associated circuitry to find the AGC bus. When you have, make the following setup.

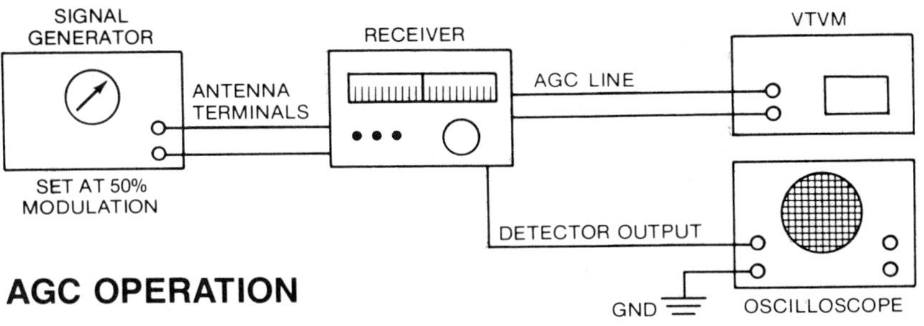

AGC OPERATION

You can observe the audio output with the oscilloscope and the AGC voltage with the voltmeter. You can vary the input-signal level from the signal generator to simulate differing signal strengths. Tune the signal generator to the receiver frequency (or vice versa). Set the signal generator to a low-level output (slightly above noise). Then, note the audio level on the scope and the AGC voltage. Record these as well as the input-signal level. Gradually increase the input-signal level and take readings until the receiver overloads. If you make a plot of the input-signal level (ordinate) versus AGC voltage and audio output (abscissa), you will get a curve that resembles the one shown below.

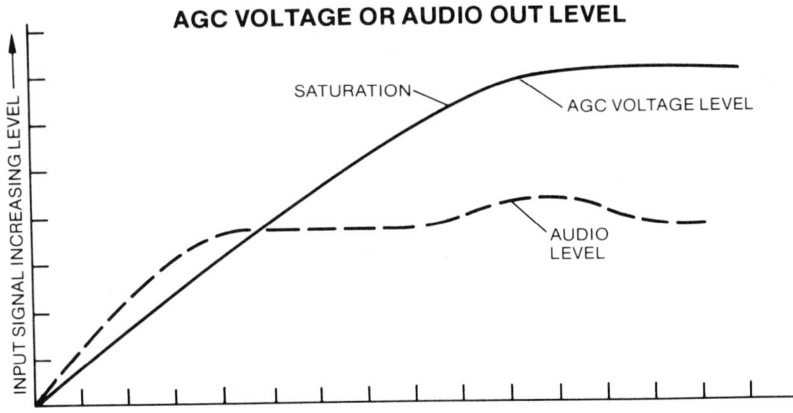

As you can see, the AGC voltage increases with increasing signal strength, but the audio output is relatively constant. You can see that the AGC action reduces the receiver gain to hold the audio output constant.

Review of Receiving Antennas and AM Receivers

1. **THE BASIC RESONANT ANTENNA** is a half wavelength long. AM broadcast receivers use loops; mobile antennas are often quarter-wave verticals.

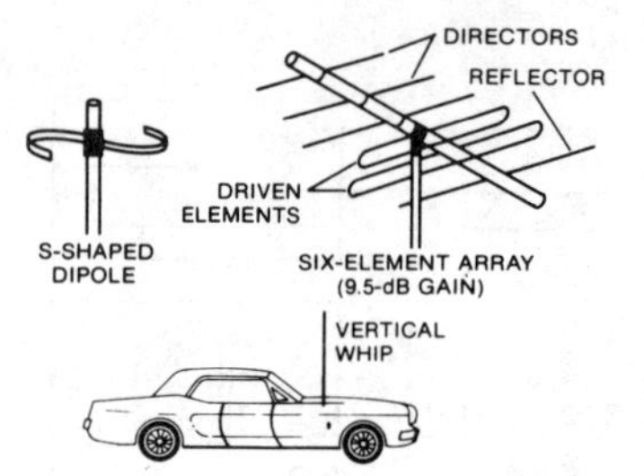

2. For FM, TV, and VHF/UHF in general, crossed dipoles, "S" dipoles, arrays, or mobile vertical whips are used.

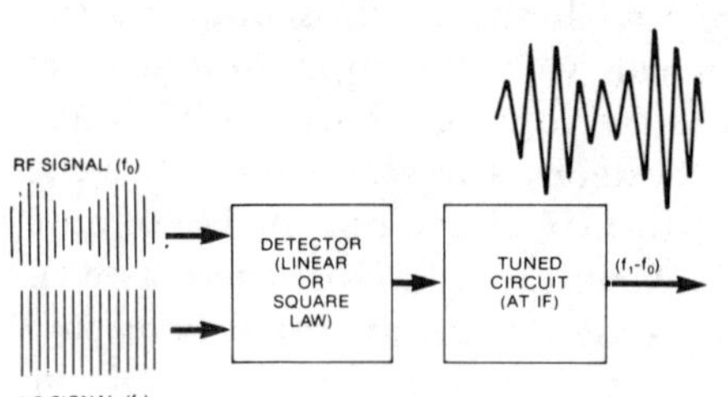

3. **RF AMPLIFIERS** precede mixers to help minimize image response. RF/mixer tuning must track with LO tuning to keep the IF signal frequency constant.

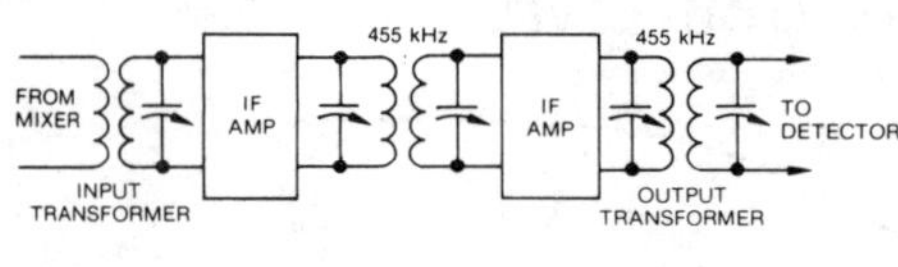

4. **THE IF AMPLIFIER** is fixed tuned to the difference between the LO and received-signal frequencies and develops practically all the selectivity and voltage amplification of the receiver.

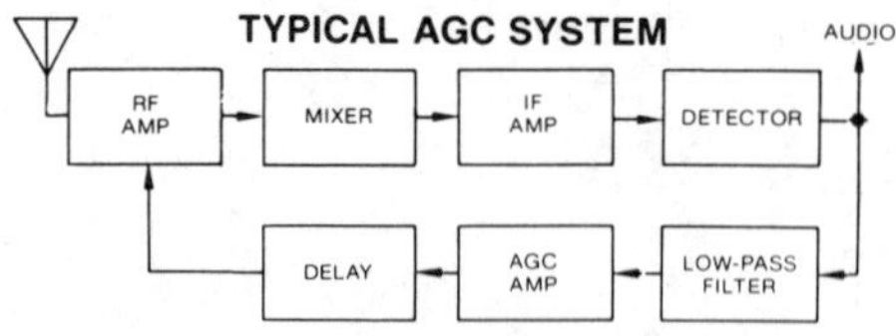

5. **AUTOMATIC GAIN CONTROL (AGC)** tends to keep the receiver output constant during changes of input signal strength. A dc voltage from the detector provides controlling bias.

Self-Test—Review Questions

1. Describe some typical receiving antennas.
2. Which do you believe make the best antennas? Why?
3. Draw a block diagram of a typical AM superhet receiver.
4. Describe the function of each block in question 3 and describe how it works.
5. Draw a diagram of an RF amplifier stage for an AM receiver.
6. Draw a diagram of a mixer/local oscillator.
7. Discuss the operation of mixers in some detail.
8. Draw a diagram of a typical IF amplifier/detector. Describe the function of the circuit elements.
9. How does an AGC work? What is it used for? What would a receiver do without AGC? Why is delayed AGC used on RF amplifiers?
10. What happens if the AGC filter has too short a time constant? Why?

Learning Objectives—Next Section

Overview—In the next section you will learn the features of FM receivers and the special circuits used to ensure good FM reception.

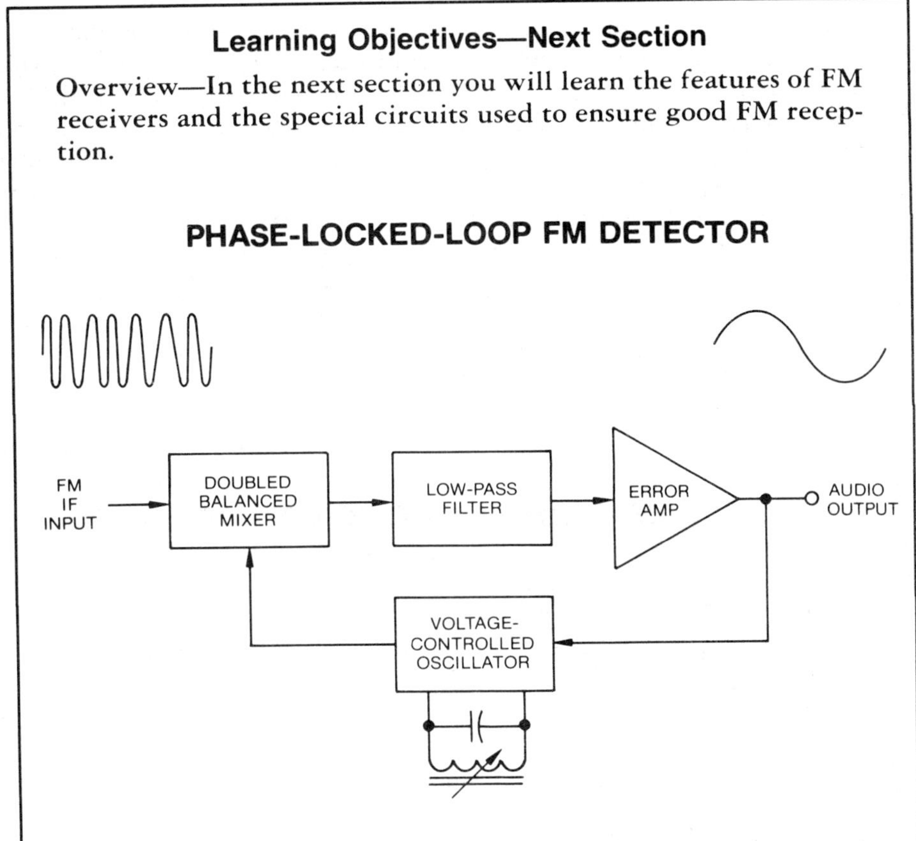

Introduction to FM Receivers

FM modulation is used in information-transmission and reception systems where *high fidelity* and *noise immunity* are desired as in FM broadcasting. These two properties are, however, obtained at the expense of *greater bandwidth* than an equivalent AM system. Because of this, FM operation is used mainly in the UHF-VHF region. FM is also used in communications systems where multipath propagation, or other interference, may cause problems. Multipath propagation takes place when the signal comes not only directly to the receiving antenna but also as the result of reflections from surrounding objects. Since multipath problems are often associated with VHF-UHF propagation, FM is used primarily in these bands. The FM broadcast band extends from 88 to 108 MHz and consists of 100 channels, each 200 kHz wide, while an FM communications band is located at 450 MHz and at other frequencies. It should be understood, however, that AM is also extensively used for point-to-point and air-to-ground communications.

You will recall from your study of transmitters and radio-wave propagation that when the signal is radiated from the antenna, part of the energy travels along the surface of the earth (known as the ground wave) and the remainder is radiated into space (sky waves). At VHF-UHF frequencies and above, the ground waves are rapidly attenuated. In addition, the sky waves are *not* reflected back toward the ground and the receiving antenna as they are at lower frequencies. Because of this, *reception* is limited approximately to *direct line of sight* between the transmitter and the receiver, and the *effective range* is limited by the *curvature of the earth*. If either or both antennas are elevated, the effective range can be extended. In cases where hills and buildings obstruct the line of sight, a severe decrease in the quality and range of reception can occur. Actually, there is some bending of the waves so that reception somewhat beyond line of sight is possible. For reasonable antenna heights, reception is typically limited to ranges of less than about 50 miles (80 km) for noise-free operation.

RECEPTION LIMITED TO LINE OF SIGHT

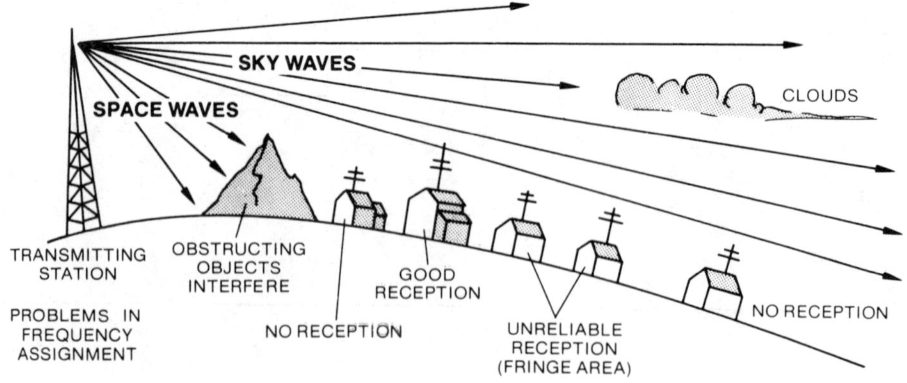

Introduction to FM Receivers (continued)

FM receivers invariably use a superheterodyne configuration, and the basic block diagram is the same as that for an AM receiver. However, the antenna, RF amplifier, and local oscillator usually operate in the VHF-UHF frequency range and require special consideration in design and construction in order to provide stable operation at these frequencies. The IF frequency for most FM sets is usually 4.5 or 10.7 MHz, and the IF amplifiers differ from their AM counterparts in a number of significant aspects that you will learn about shortly.

In both AM and FM receivers, the purpose of the stage following the IF amplifier is to extract the information from the carrier signal. Since the information was originally impressed on the carrier by modulating its frequency, the FM detector must sense *frequency*—not amplitude—variations. Thus, the basic differences between an FM and AM receiver lie in the *detector circuit*. Also, since FM signals may occupy a wider band for a given modulating signal, the *RF and IF bandwidths are typically greater*. Since FM is usually confined to higher frequencies, this wider bandwidth requirement is more tolerable.

The audio portion of an FM receiver is identical to that for an AM receiver, except that the audio portion of an FM broadcast receiver is usually capable of amplifying without distortion a wider frequency range than an AM broadcast receiver—typically 15 kHz versus 5 kHz.

In the sections that follow, you will learn about the performance characteristics of each of the circuits that are of importance to FM reception. Special emphasis and detailed study will be devoted to the IF amplifier and detector sections—the heart of the FM receiver and the only circuits that may be significantly different from an AM receiver. While the principles described here are applicable to any FM receiver, emphasis will be on FM *broadcast* receivers as a mechanism for studying FM receivers.

FM RECEIVER

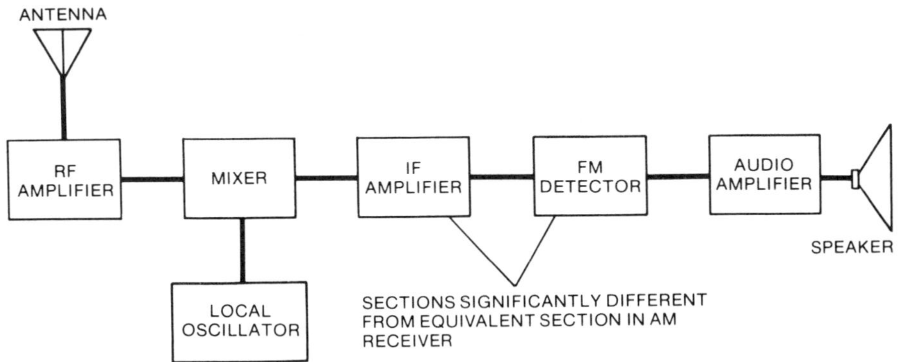

Introduction to FM Receiver Circuits

As an introduction to FM receivers, a complete FM broadcast receiver will be described on a stage-by-stage basis. First, a stage or group of stages will be described in terms of discrete, transistorized elements. Then, when appropriate, you will learn how these same stages have been designed in more advanced solid-state elements such as MOSFETs and ICs. The FM section of an AM/FM receiver will be used as the learning vehicle for the discrete transistor discussion. You will notice that the block diagram of this receiver looks very much like an AM broadcast receiver, and the only new elements are the limiter (Q8), and the discriminator (CR19–CR20). Thus, as you can see, the only basic difference between an FM and AM receiver involves the detector. In one case, the detector produces an output signal proportional to the *carrier amplitude*; in the other case, the output signal is proportional to the *frequency deviation* of the received signal.

As you have already learned, the FM broadcast band in the United States consists of 100 channels (channel numbers 201 to 301) covering the frequency range from 88 to 108 MHz. This is the frequency over which the RF amplifier's and mixer's input must tune. The first channel (201) is *centered* at 88.1 MHz and the last channel (301) is *centered* at 107.9 MHz. Since the IF is 10.7 MHz, the oscillator (usually tuned to the high side) must tune from 98.8 to 118.6 MHz.

FM SECTION: TYPICAL FM RECEIVER

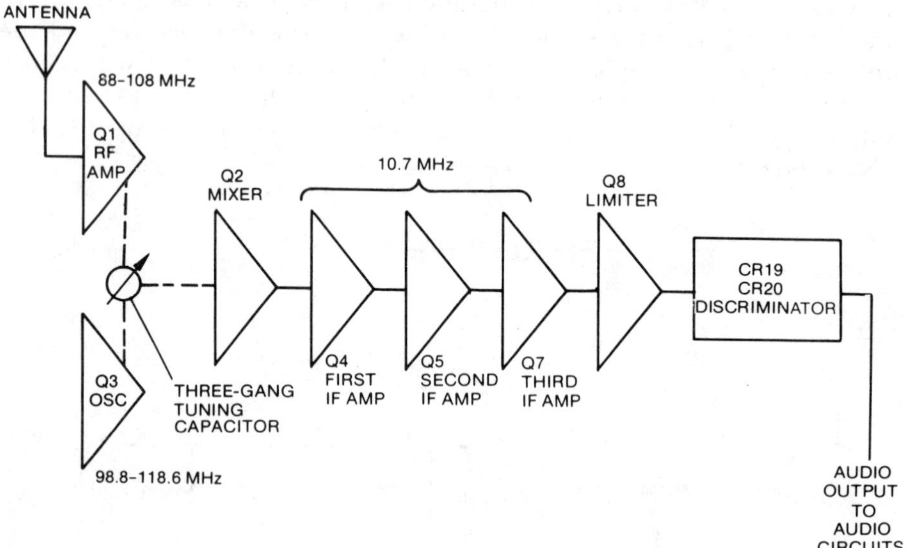

RF Amplifier, Mixer, and Local Oscillator

The FM RF circuits are tuned with a three-gang tuning capacitor (C1), which is coupled to the tuning dial. A fixed capacitor and a small trimmer capacitor are connected in parallel with each of the three sections to provide for proper alignment (see next page for schematic).

The FM signal at the antenna terminals, normally in the microvolt range, is coupled to the base of Q1 by input transformer T1. The input signal is amplified in RF amplifier Q1 and then coupled through C11 to the base of the mixer (Q2). A tuned circuit comprised of C1 (B) and inductor L2 provides for tuning of the RF amplifier mixer interstage. The local-oscillator signal is also coupled to the base of the mixer through C12. The oscillator stage, Q3, is tuned by one section of C1 and inductor L4. The oscillator coil has a movable core to permit oscillator alignment with the tuning dial.

The mixer stage is biased near cutoff by R8 and R9, and driven into conduction during the negative peaks of the oscillator signal. The RF signal, also present at the base of Q2, alternately adds to and subtracts from the oscillator signal in the mixer stage. The output of the mixer contains, in addition to the original RF and oscillator signals, the sum and difference frequencies. The desired difference frequency of 10.7 MHz is coupled through fixed-tuned transformer L5 to the first IF amplifier. The 100-MHz (nominal) input signals and the 200-MHz sum signal are shunted to ground through the transformer primary network and C19. As you can see, except for component values that reflect the operation in the VHF range, the circuits are essentially identical to those for AM. C14 in the LO tuned circuit has a temperature characteristic chosen so that mistuning does not occur as the temperature changes.

The development of MOSFET devices has made it possible to make considerable improvements in FM receiver performance. MOSFET devices are used in RF stages in preference to bipolar transistors because of their greater dynamic range, lower circuit noise, and higher gain. The primary improvements are their better performance due to less noise and ability to handle a greater range of input-signal levels. Conventional bipolar transistors show relatively larger change in input capacitance as a function of signal level, which can cause circuit detuning at higher frequencies; these effects are absent in MOSFETs. There is little loading of the input signal or drastic change in input capacitance even with extremely large input signals. The dynamic range of the MOSFET is about 25 times greater than bipolar transistors, but the differences in circuit source impedances reduce this to a practical value of about five times. The linearity of MOSFET RF amplifiers therefore ensures good performance at high signal levels.

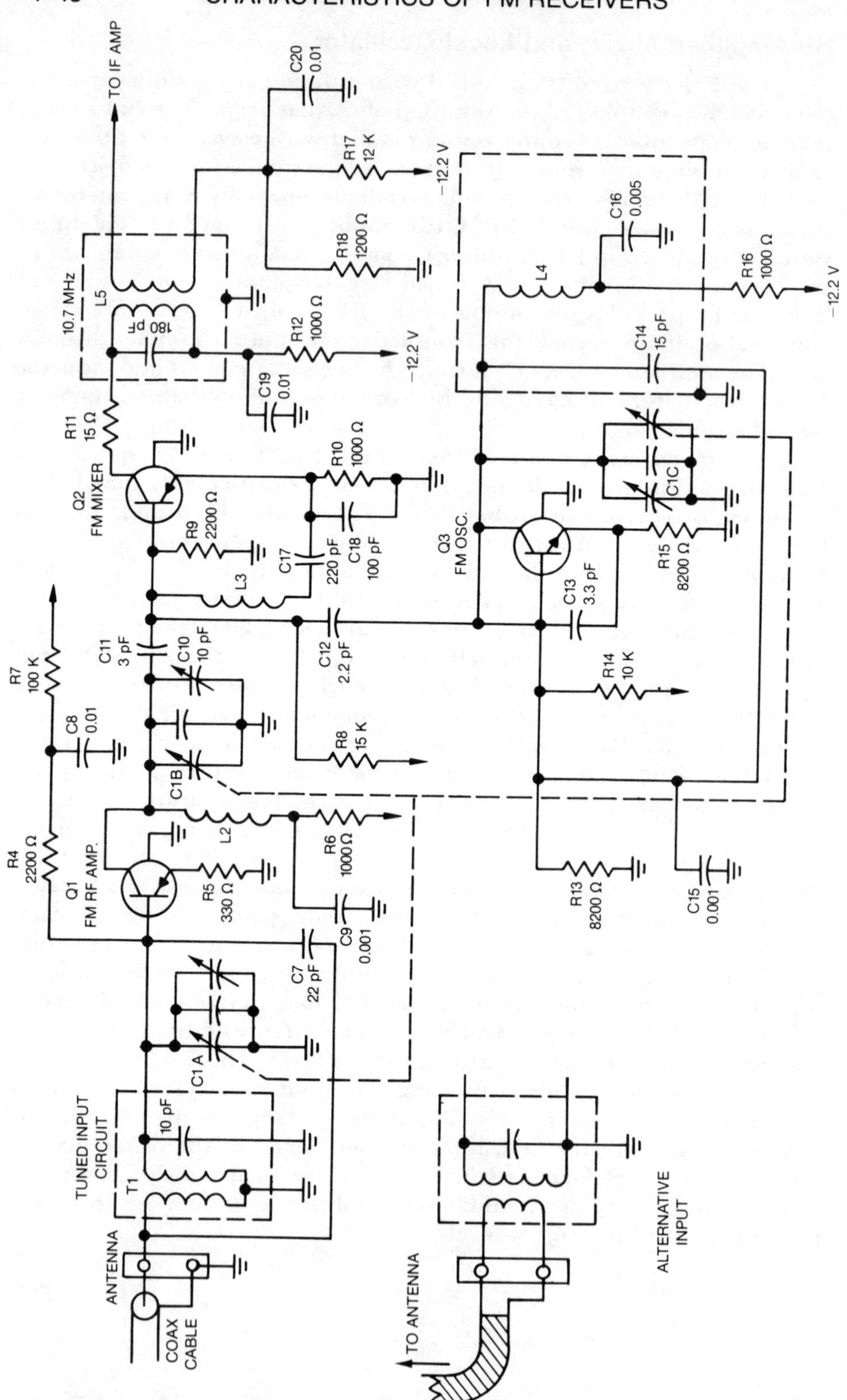
TO IF AMP
C20
0.01
R17
12 K
−12.2 V
R18
1200 Ω
10.7 MHz
L5
180 pF
R12
1000 Ω
−12.2 V
C19
0.01
R11
15 Ω
Q2
FM MIXER
R10
1000 Ω
R9
2200 Ω
C17
220 pF
C18
100 pF
L3
C11
3 pF
C10
10 pF
C12
2.2 pF
R7
100 K
C8
0.01
R8
15 K
C1B
R4
2200 Ω
L2
R6
1000 Ω
Q1
FM RF AMP.
R5
330 Ω
C9
0.001
C7
22 pF
C1 A
TUNED INPUT
CIRCUIT
10 pF
T1
ANTENNA
COAX
CABLE
TO ANTENNA
ALTERNATIVE
INPUT
C16
0.005
L4
R16
1000 Ω
−12.2 V
C14
15 pF
C1C
Q3
FM OSC.
R15
8200 Ω
C13
3.3 pF
R14
10 K
R13
8200 Ω
C15
0.001

FM IF Amplifiers

In an AM broadcast receiver, the IF amplifiers are synchronously tuned (all tuned to the same frequency), with the result that the IF passband has a single peak and the tuned-circuit rolloff provides good selectivity. Since the maximum AM audio frequency is typically about 5 kHz, the double-sideband AM spectrum occupies 10 kHz. If the bandpass is just 10 kHz wide, the high-frequency (5-kHz) audio components are passed with 3 dB less gain than low-frequency components. While this results in some loss of fidelity, little harmonic distortion results.

The situation with FM IF amplifiers is quite different. An FM broadcast station is permitted a peak deviation of 75 kHz (150 kHz peak-to-peak). For FM reception, a more nearly rectangular passband than for AM is desired for two reasons. The first is harmonic distortion and phase shift of the relative frequency components; the second is channel assignment and selectivity.

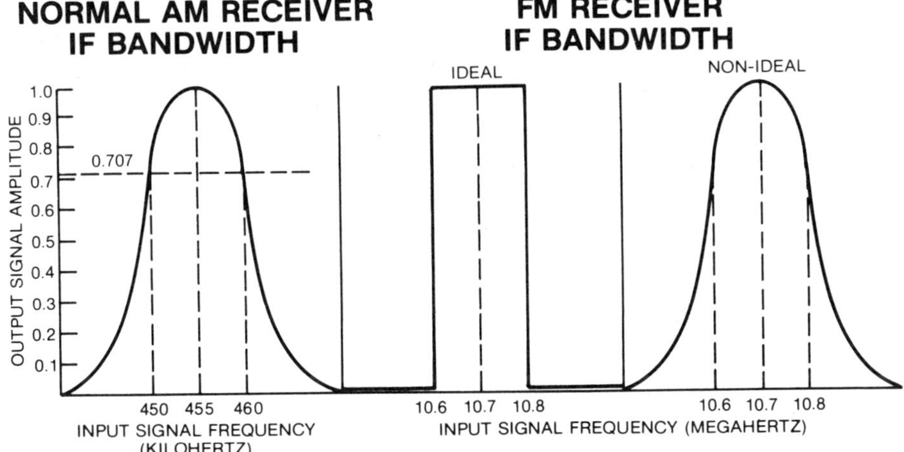

You will recall that FM deviation is a function of the amplitude or loudness of an audio signal and *not* its frequency. Therefore, if a loud tone, whether high or low frequency, is passed through an IF amplifier with limited frequency response, the result is a clipping or rounding of the peak of the audio waveform, causing harmonic distortion. The second factor, requiring a flat IF frequency response with sharp out-of-band rolloff, is the need for good adjacent channel rejection in FM receivers. Because of the line-of-sight limitation in the FM broadcast band, it is very possible to have large differences in signal strength which are *not* functions of the distance from the transmitting station but of relative antenna altitudes. Therefore, while the FCC limits channel assignments to minimize interference, it is entirely possible for a distant station to have a stronger signal at the user's location than that of a local station being listened to. If selectivity is not sufficient to attenuate this out-of-band station, its signal will affect the action of the limiter and can reduce, or even completely suppress, the signal from the local station. FM receivers therefore are designed to approximate the ideal response.

FM IF Amplifiers (continued)

Three methods are generally used to provide the desired FM IF response curve. The first method, known as *stagger tuning,* gives an excellent frequency-response curve but reduced gain. Three IF amplifier stages are tuned, for example, to 10.6, 10.7, and 10.8 MHz, respectively. The overall frequency response of this combination adds to produce the close-to-ideal response curve shown in the diagram below. The overall gain of the three stages is not much greater than that of a single stage. In addition, special alignment techniques are required. As a result, this method is not used as frequently as other methods for FM, but is frequently used for other wide-band IF applications such as TV and pulse receivers.

The second method of approximating the desired broad frequency response is to use three IF amplifiers, all tuned to the same center frequency, synchronously tuned. A broader response is obtained in each stage by using less-than-critical coupling in the transformers and by using transformer coils with low Q or by deliberately adding resistance across the tuned circuits to reduce the Q. This method does not result in very good selectivity (see the top figure on page 4–51).

STAGGER TUNING

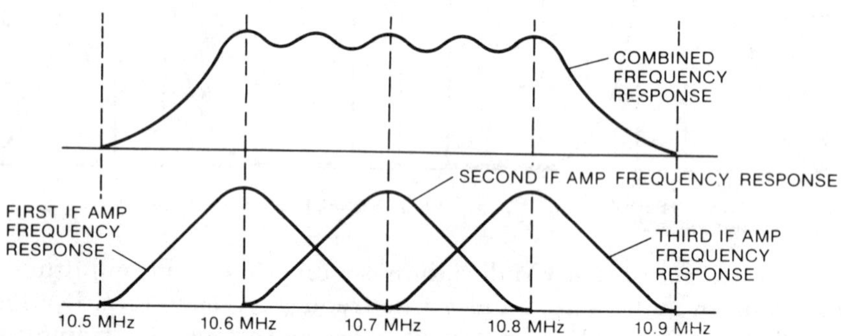

The third method also makes use of two or three IF tuned circuits tuned to the same center frequency. The first and third stages are designed and tuned as in the second method noted above. However, the transformer of the second stage is overcoupled, resulting in the familiar double-peaked frequency-response curve. When the individual response curves of the three stages are combined, the double peaks of the second stage have a significant effect in broadening the frequency response and steepening the skirts, producing a good approximation of the desired response curve. The good response characteristics obtained make this method a popular one (see the bottom figure on page 4–51).

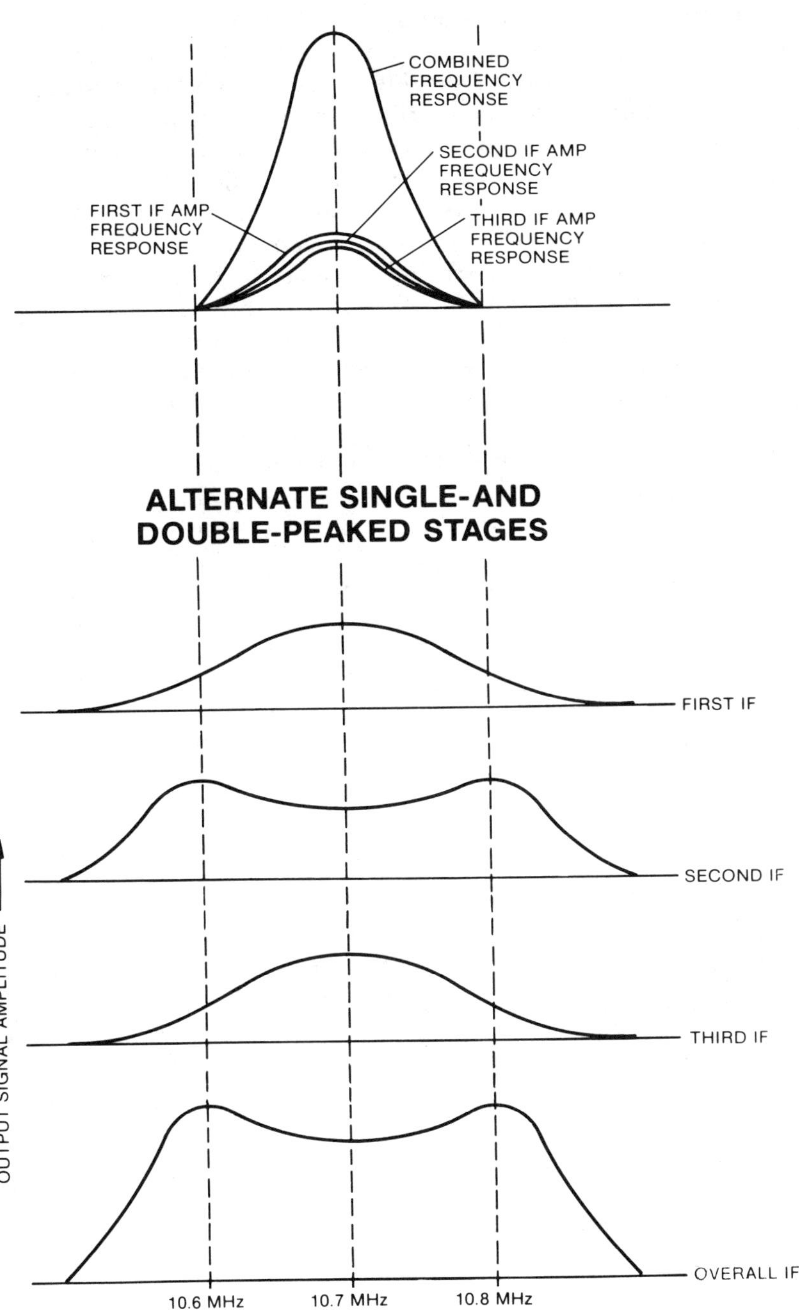
THREE IFs WITH SAME CENTER FREQUENCY
COMBINED FREQUENCY RESPONSE
SECOND IF AMP FREQUENCY RESPONSE
FIRST IF AMP FREQUENCY RESPONSE
THIRD IF AMP FREQUENCY RESPONSE
ALTERNATE SINGLE-AND DOUBLE-PEAKED STAGES
FIRST IF
SECOND IF
THIRD IF
OVERALL IF
OUTPUT SIGNAL AMPLITUDE
10.6 MHz
10.7 MHz
10.8 MHz

Ceramic Filters

Although electromechanical and crystal filters have been employed for many years, the advent of microelectronic circuits has resulted in a proliferation of the equipment and technology needed to produce low-cost ceramic filters. Nowadays, all high-quality FM broadcast receivers use these filters to provide good selectivity and passband shape. These filters are based on the piezoelectric effect in the ceramic material, and have the inherently very high Q normally associated with quartz crystals.

Ceramic filters can range from simple single-element devices to complex multiple-element units. However, even multiple-element units are small in size at 10.7 MHz. If a single-element, two-terminal filter is placed across the emitter resistor of an IF amplifier the amplifier has a gain about equal to the transistor β at f_o, the series-resonant frequency of the crystal element, because the series-resonant crystal essentially grounds the emitter at the intermediate frequency. At other frequencies, the gain is approximately R_L/R_E, which can be quite low if R_E is large.

CERAMIC FILTERS

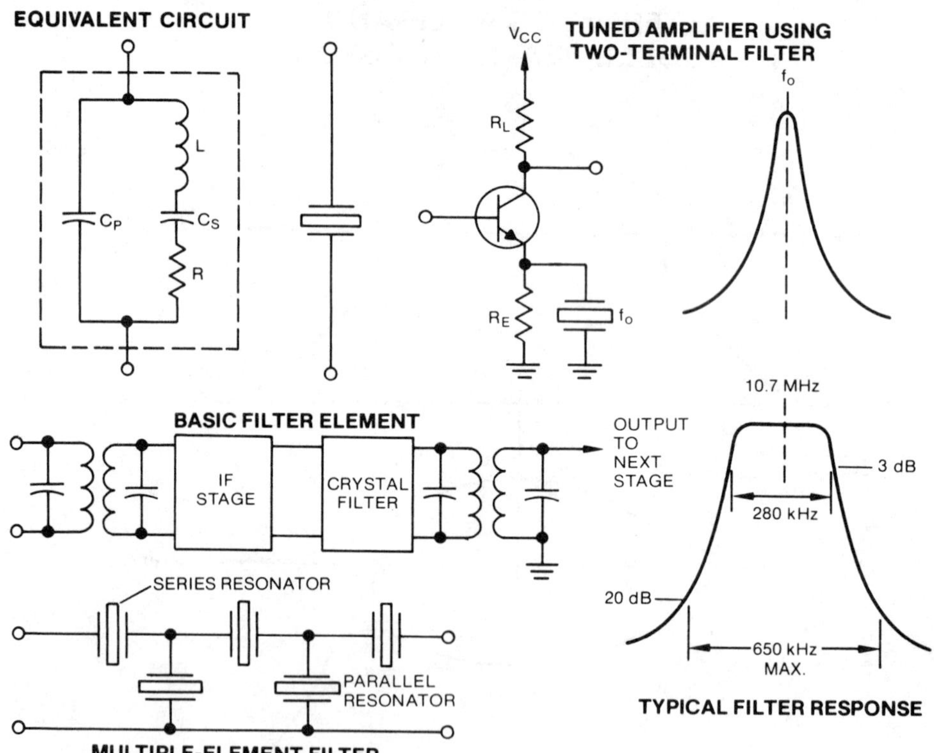

Ceramic Filters (continued)/IF Amplifier Circuit

When two or more elements are connected in a four-terminal arrangement and used between IF stages, substantial increases in bandwidth and selectivity can be obtained. Also, the input and output impedances can be controlled, and made unequal if necessary for interstage matching purposes. By controlling the coefficient of electromechanical coupling between filter elements, the bandwidth can be very accurately shaped, both in width and flatness. Ceramic filters are widely used in communication equipment and in FM receivers to provide high selectivity to avoid adjacent channel interference.

Now let us move on to our next subject—the study of the IF amplifier circuit.

The example of the IF amplifier shown on page 4–54 is part of the same receiver previously described with transistor-type RF amplifier, oscillator, and mixer. This receiver has three IF stages consisting of Q4, Q5, and Q6. The input and output of each stage is transformer-coupled with a standard FM tuned IF transformer at 10.7 MHz. Tapped transformer windings are used to provide better impedance matching, and coupling is adjusted to provide an optimal passband shape.

The signal from the mixer is coupled to the base of Q4 via an input transformer, and Q4 output is coupled through R20 to the second IF transformer L6. Both the primary and secondary windings of L6 are inductively tuned to the FM intermediate frequency of 10.7 MHz. The signal in the secondary winding of L6 is applied from the winding tap to the base of the second IF amplifier (Q5), where it is further amplified and coupled through R27 to another double inductively tuned IF transformer (L11), which is identical to L6.

From the secondary of L11, the amplified signal is supplied directly to the base of the third FM IF amplifier (Q6). This stage is the last amplifier stage for the IF signal. After amplification, the signal is again coupled through resistor R32 to a double inductively tuned transformer L13. The IF signal is then applied to the amplitude limiter stage Q8, to be described later.

The bases of the three IF amplifiers are biased at about -1 volt by resistive dividers and decoupling networks similar to those shown for the RF amplifier and mixer stages. Resistors R20, R27, and R32 are used to provide suppression of undesired oscillation in the IF amplifier stages because of inadvertent feedback. The IF stages shown are neutralized to minimize instability and to minimize tuning changes as a result of change in input-signal levels.

After you review the schematic on the next page you will be introduced to the three basic FM detectors and their properties.

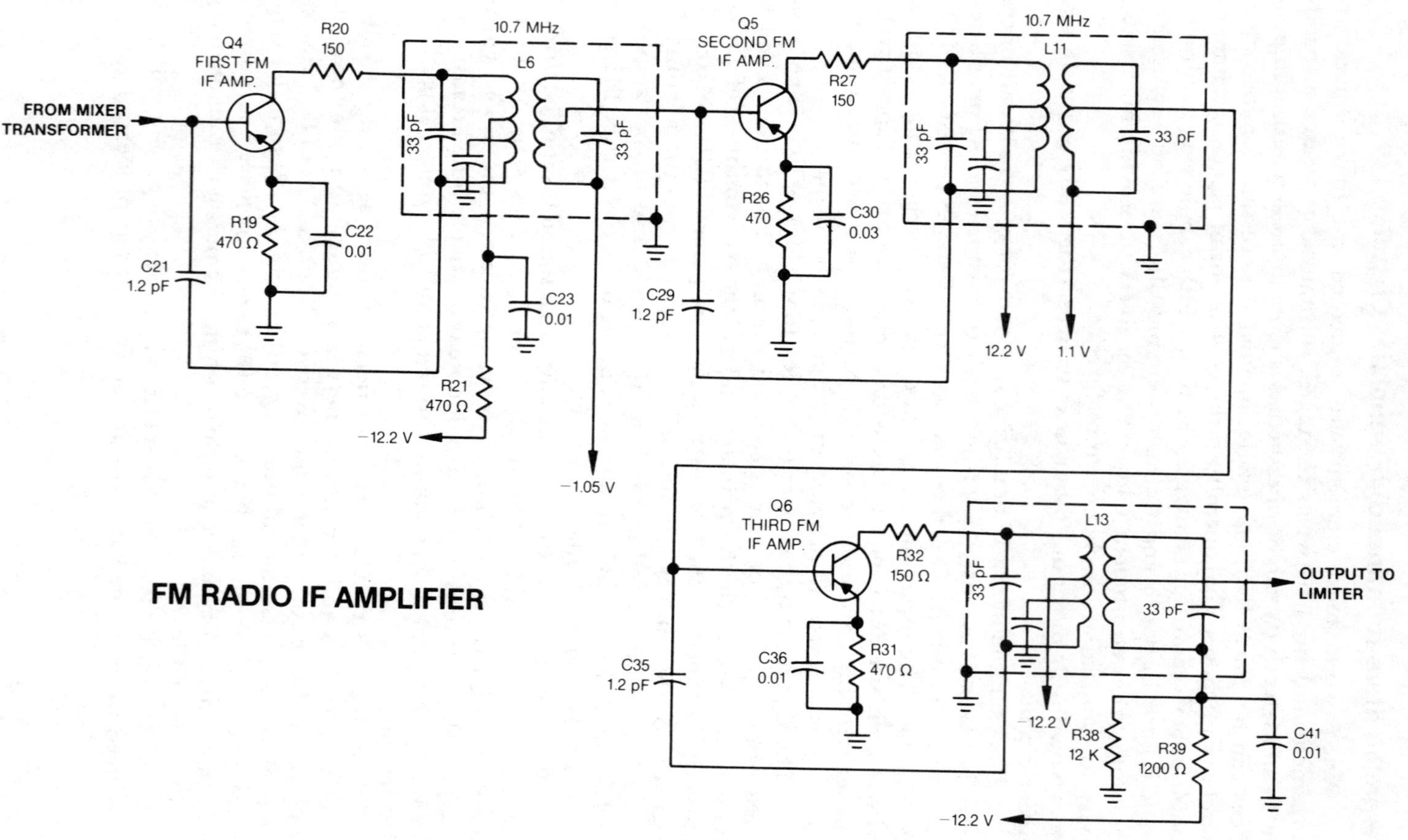

FM RADIO IF AMPLIFIER

FM Detectors—Limiting

In your earlier study of the use of frequency modulation for the transmission of information, you learned that the carrier-signal amplitude did not carry any information and that all intelligence was contained in the frequency variations of the RF carrier. Because of these factors, FM detectors are quite different from their AM counterparts and require completely new circuits.

These basic detector types are commonly used—the *slope-detector*, *phase-detector*, and *frequency-following* (or *phase-locked-loop*) detectors. Some of these detectors are sensitive to amplitude variations and, as you will learn shortly, some are not. If the FM receiver uses an amplitude-sensitive FM detector, then the detector stage must be preceded by a *limiter* to remove from the signal any amplitude variations that occur from interference signals caused by lightning flashes, neon signs, automobile ignition, microwave ovens, small appliances, and various other spark-producing electrical equipment. The limiter removes the signal-amplitude variations caused by these disturbances but leaves the original FM intelligence unaltered, since the frequency is not changed by limiting. It is the *limiter*—either as a separate stage or as the absence of amplitude sensitivity in the detector—that gives the FM receiver its well-known freedom from noise and immunity to interference from other signals. This immunity occurs because the weaker signal added to the stronger signal is stripped off by the limiter and does not appear at the limiter output.

LIMITER AND DISCRIMINATOR FUNCTIONS

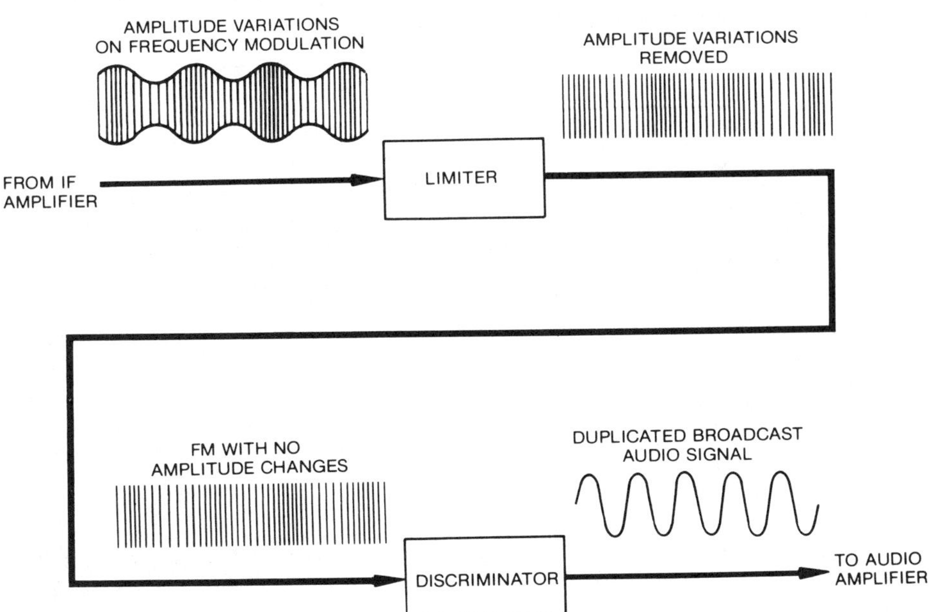

FM Detectors—Limiting (continued)

Limiters use saturation and cutoff to limit the positive and negative signal peaks. As shown below, operation is linear for small signals; but as the signal increases in amplitude at the input, the output does not increase because of limiting action. The limiter responds to the amplitude variations for very small signals (1); however, for higher level signals (2), the signal output remains constant.

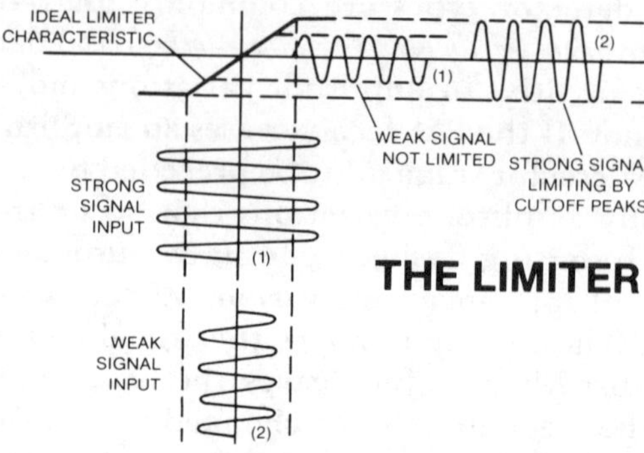

THE LIMITER

There are many circuit configurations for limiters. Basically, these depend on cutoff and/or saturation of diodes or transistors. The simplest limiter is a pair of back-to-back diodes. As you remember, there is a minimum voltage below which the forward bias is insufficient to narrow the depletion layer gap width, so that the diode does not conduct. If a pair of diodes is connected as shown, the input signal is unaffected by the diodes, but the diodes conduct (CR1 for positive peaks and CR2 for negative peaks) so that clipping or limiting occurs for signals large enough to overcome the offset voltage of the diode.

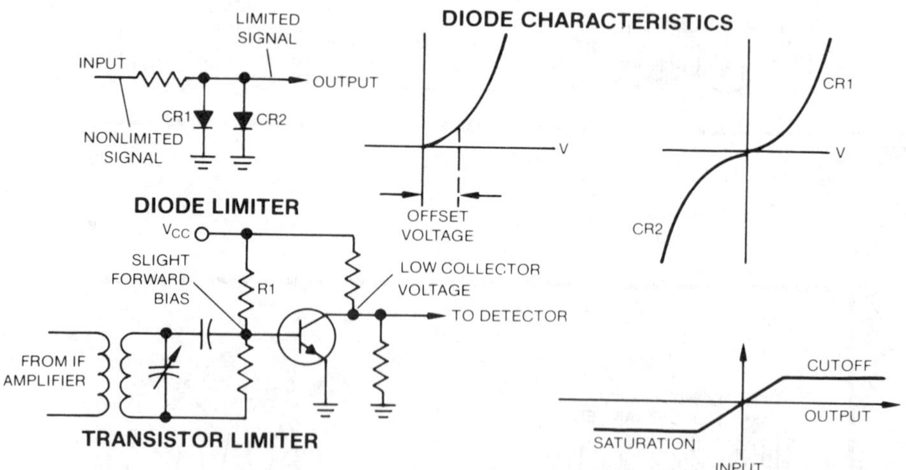

The transistor limiter is biased on by the forward bias through R1 and is driven to cutoff by relatively small negative signals. By keeping the collector voltage very low (1.5–2 volts), saturation occurs with relatively small positive signals. In this way, limiting action takes place.

FM Detectors—Slope Detectors

The simplest FM detector is the slope detector that uses one resonant circuit tuned to one side of the carrier, converting frequency deviation to an amplitude variation because of the unbalanced frequency response. This amplitude variation can then be detected by a simple AM (diode) detector. Because of its sensitivity to carrier detuning/drift, this simple detector has little practical application, and the *balanced slope detector* (or *diode discriminator*) is used instead. This configuration uses two slope detectors, one tuned to either side of the carrier frequency.

In the circuit shown on the next page, the final IF transformer has a center-tapped secondary. The primary is tuned to the IF frequency (10.7 MHz for FM broadcast) and the two halves of the secondary (L_x–C_x and L_y–C_y) are tuned, for example, to 10.6 and 10.8 MHz, respectively. If no modulation exists on the FM carrier, the IF frequency is at 10.7 HMz, and the detected currents through R_x and R_y are equal and opposite, so the voltage across R_L (equal and opposite currents) is zero. If no modulation exists but the local oscillator is detuned, the IF is not at 10.7 MHz, and a steady-state dc error voltage is produced across R_L because I_x does not equal I_y. This error voltage can be used to correct the oscillator frequency (automatic frequency control, or AFC), as will be described later. Unfortunately, the slopes of the two tuned circuits are neither linear nor constant, so although the circuit shown will work, it is rarely used in a practical FM receiver. It is important to understand how the slope detector works, since this helps in understanding more widely used FM detectors.

OPERATION OF BASIC DIODE SLOPE DETECTOR

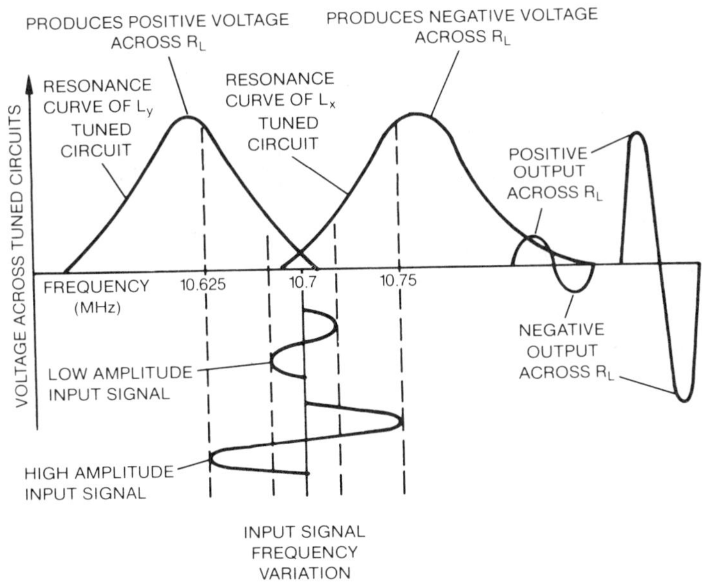

FM Detectors—Slope Detectors (continued)

Let us assume that the RF carrier has been frequency modulated by a low-deviation (low-amplitude) 1,000-Hz tone to produce a total carrier deviation of ±15 kHz. The IF signal will then deviate 15 kHz to each side of the IF frequency at a 1,000-Hz rate. When the IF signal is 15 kHz higher than 10.7 MHz, more current will flow through R_x and a smaller current through R_y. The net result is an imbalance of current through R_L and a positive voltage at the output. Now, as the IF signal swings back across 10.7 MHz—following the instantaneous envelope of the 1,000-Hz modulation—an increasingly larger current flows through R_y and a decrease occurs in the current through R_x. Thus, the current through R_L, which is the sum of the currents, will produce a negative voltage at the output. Thus, the output follows the variations in the IF frequency. Since the IF frequency was changed at a 1,000-Hz rate, a 1,000-Hz signal appears at the audio output.

If the frequency deviation increases, as it would with a louder 1,000-Hz tone, larger difference current will flow through R_L and a large 1,000-Hz voltage appears at the audio output.

You will note that when no modulation is applied the voltage across R_L is zero, and therefore this detector is not sensitive to variations in carrier amplitude. A fixed-frequency offset will produce a dc voltage (+ or −), depending on how far the frequency is offset. This dc voltage can be used to provide automatic frequency control. However, when modulation is applied, the output is proportional to the amplitude of the input signal as well as its frequency. Therefore, an IF limiter is required when a slope detector is employed. Sometimes this detector is called a *high/low FM detector*. It is not usually used in FM broadcast receivers because it is not very linear, and hence distortion results. It is mainly used in special broadband applications in radar receivers.

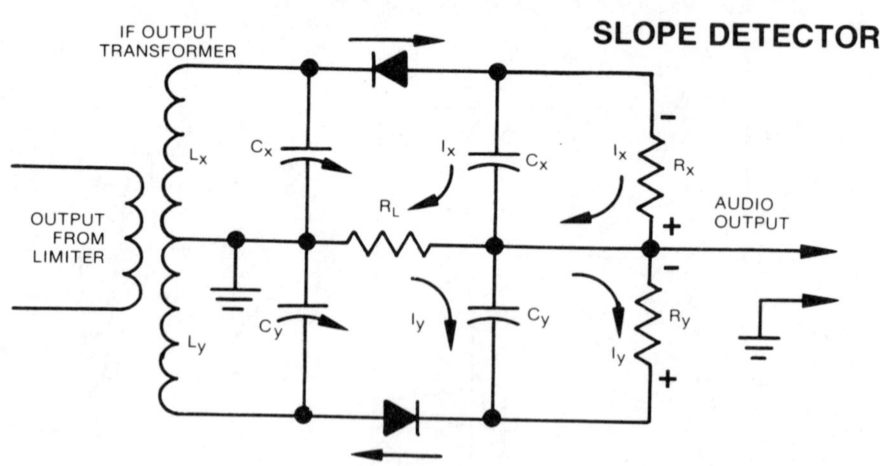

FM Detectors—Quadrature Detectors

An important group of PM detectors is based on the combining of *quadrature* voltages (signals that are 90 degrees or $\pi/2$ radians out of phase). You learned during your study of transmitters how quadrature mixers were used to generate both SSB and FM signals.

Assume two signals (E_a and E_b) are 90 degrees out of phase. The 90-degree phase shift in E_b is produced either by capacitor C or inductor L. If these two signals are added vectorially, the resultant is as shown in diagram A. Since the secondary of the transformer is tuned, the output at E_a shifts in phase relative to E_b above and below resonance. As you remember, above resonance X_C is less than X_L and the parallel circuit looks capacitive; below resonance X_C is greater than X_L and the parallel circuit looks inductive. Therefore, as the FM signal varies in frequency, the phase of E_a shifts relative to E_b, as shown in diagrams B and C. This phase shift (ϕ) is proportional to the FM deviation, as shown. If these two signals are added, properly detected, and then filtered to remove all carrier-frequency components and higher harmonics, the output is a signal proportional to $E_aE_b \cos \phi$. Therefore, FM detection can be accomplished by providing two signals which are in quadrature when the IF (or RF) signal is at its center frequency of 10.7 MHz and which deviates from this phase relationship as the frequency varies. This principle forms the basis for two of the most common FM detectors, the *Foster-Seeley discriminator* and the *ratio detector.*

QUADRATURE DETECTOR

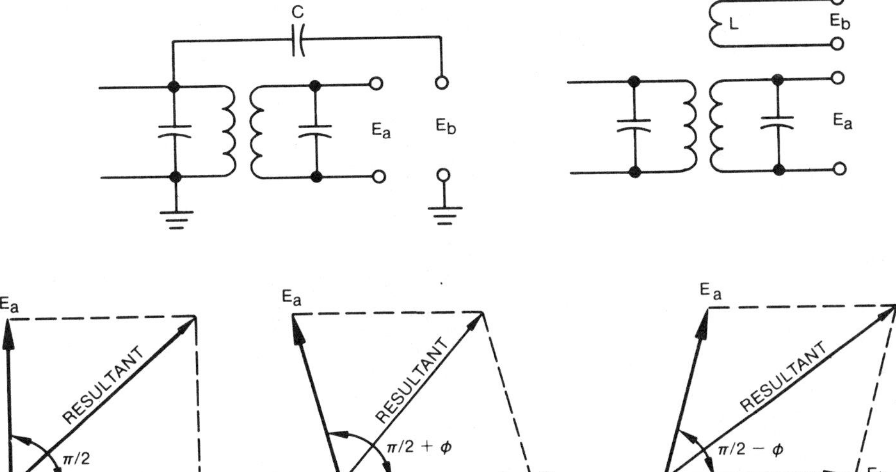

FM Detectors—Foster-Seeley Discriminator

In the Foster-Seeley discriminator bypass capacitors C1 and C2 maintain the cathodes CR1 and CR2 and the end of RF choke L at ground, but are too small to bypass the audio signals. Voltage E_p appears across the choke, and the secondary voltage appears equally across each half, shown here as E_A and E_B (see A).

Voltages E_A and E_B are equal but 180 degrees out of phase with respect to the secondary center tap. The primary voltage E_p is coupled through capacitor C and appears 90 degrees out of phase with the total secondary voltage across L. Voltages E_p and E_A add to produce E_1, and voltages E_p and E_B add to produce E_2. The rectified currents I_1 and I_2 flow in opposite directions and the voltages across R1 and R2 are *equal and opposite* at resonance (see A). Thus, the audio output voltage is zero when $R_1 = R_2$ and $E_1 = E_2$. This reduces noise and AM. When E_1 exceeds E_2 (above resonance), the output voltage is positive (B). When the carrier swings below resonance, E_2 exceeds E_1 and the audio output is negative (C). The audio output circuit is usually taken from the top end of R1, as shown. Since any offset in frequency appears at the junction of R1–R2 as a dc voltage, AFC voltage is also available at this point. This discriminator is sensitive to AM signals off center frequency and therefore requires a limiter.

FOSTER-SEELEY DISCRIMINATOR

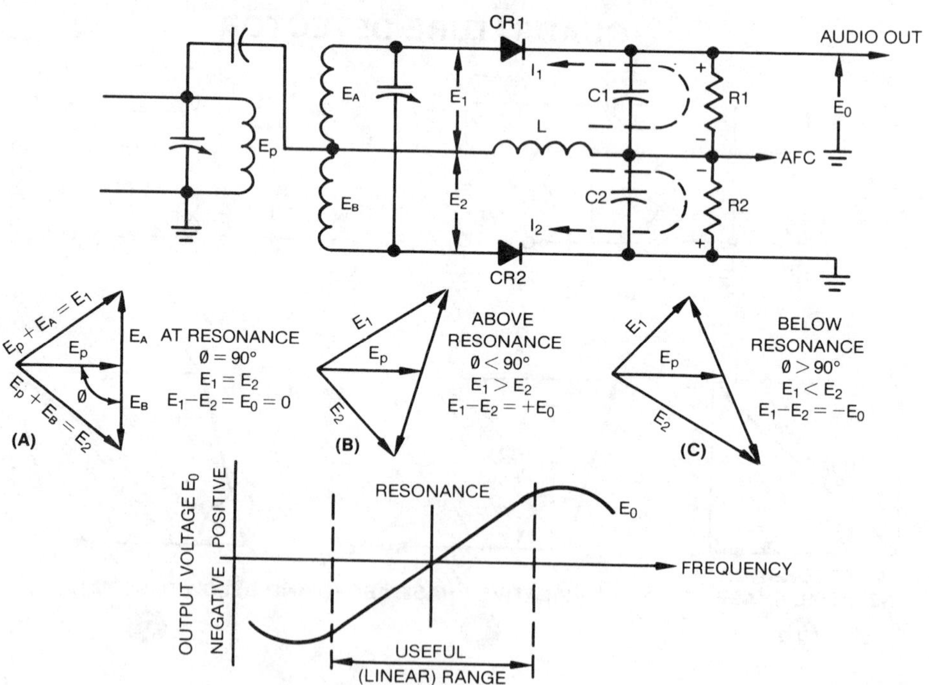

FM Detectors—Ratio Detector

The ratio detector is a quadrature detector that is relatively insensitive to short-term amplitude variations in the carrier. Therefore, a separate limiter is *not* needed, and this circuit is very popular in FM receivers. You will note that the ratio detector circuit is very similar to the Foster-Seeley discriminator with the principal change being the reversal of one of the diodes and the addition of C3. Because of the way the diodes are connected, the total voltage (sum) across C1 and C2 is constant for a given carrier level. However, the difference (at the junction of C1 and C2) is dependent on the instantaneous frequency—or audio signal. Thus, an audio output can be taken from this point. The time constant for R1 + R3 and C3 is long compared to the lowest audio frequency of interest (typically 0.1 second), so that the total voltage across R1 + R2 can be changed only slowly.

If the input-carrier level should shift slowly, the voltage across C3 will readjust to the new carrier level. However, if the voltage shifts abruptly, the voltage across C3 cannot shift abruptly, since current is required to charge (or discharge) C3. Thus, with rapid carrier shifts due to noise, part of the current must be supplied to C3, and this current is diverted from across R1–R2 into C3. Conversely, if the carrier abruptly shifts downward, less current goes into C3 and more goes through R1–R2. Thus, the total voltage across R1 and R2 is held constant. Long-term changes in carrier level—as between stations, or for slow fading, allow the average voltage across C3 to vary slowly, and hence the peak-to-peak audio output voltage from a ratio detector is not only proportional to the FM deviation but also the average carrier level.

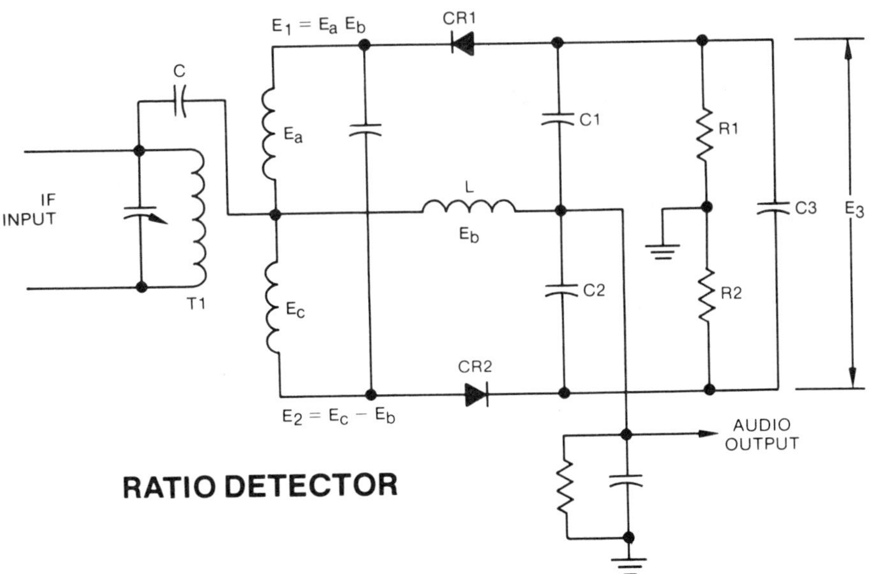

RATIO DETECTOR

FM Detectors—Phase-Locked Loop

The PLL (phase-locked-loop) detector is a variation of the PLL you learned about in your study of transmitters. Because of its freedom from tuning elements and the ease with which it can be implemented in IC form, the PLL is very popular for FM receivers. Most PLL ICs use a double-balanced mixer as the phase detector since the double-balanced mixer suppresses level changes. Thus, the PLL FM detector needs no limiter. The FM input signal (10.7 MHz) is mixed with the output of the voltage-controlled oscillator (VCO), which is at a nominal frequency of 10.7 MHz also. The mixer output contains both the sum and difference frequencies—21.4 MHz and the VCO tuning error (difference signal). The sum signal is removed by the low-pass filter, and the difference (error) signal is amplified and drives the VCO into frequency and phase coincidence with IF input.

The input IF signal changes frequency in accordance with the information content and the PLL amplifier develops a signal equal and opposite to the original modulation that keeps the VCO tuned to the instantaneous IF frequency. Thus, the output of the error amplifier is a signal that duplicates the original frequency modulation impressed on the carrier back at the FM transmitter.

Since the PLL frequency discriminator or tracker contains no tuning elements, the total circuitry is readily implemented in IC form and has found broad application in modern FM systems. Furthermore, since the VCO error voltage (from which the audio output is derived) is proportional to frequency only, no limiter is needed.

On the next page we will evaluate and compare the Foster-Seeley discriminator and limiter, the ratio detector, and the phase-locked-loop detector. Included is a chart for easy comparison of the three detectors.

PHASE-LOCKED-LOOP FM DETECTOR

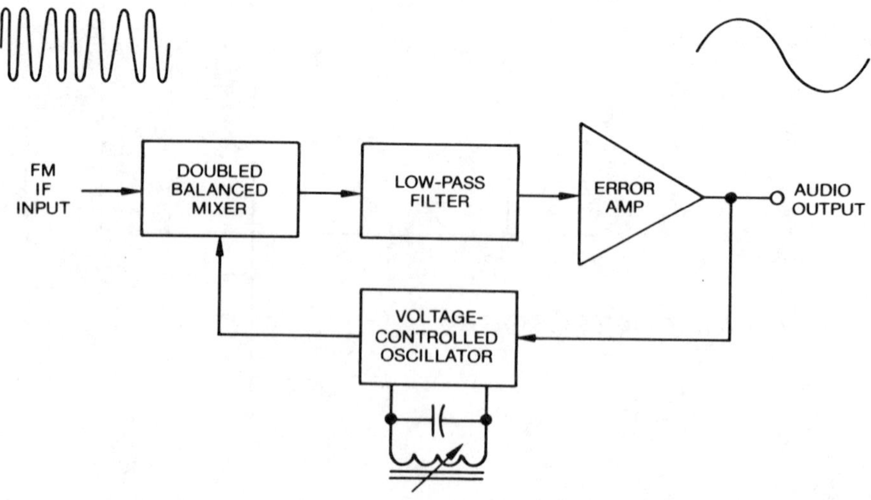

Evaluation of FM Detectors

Both the limiter-discriminator and ratio detector arrangements have advantages and disadvantages when compared with each other. Because of the nature of the relative advantages and disadvantages, there can be no clear-cut statement of which is better. Each type is preferred by some designers, and it is essentially a matter of economics as to which is used.

The important advantage of the limiter-discriminator arrangement is that it is relatively a simple matter to balance the two sides of the discriminator and obtain excellent reproduction of the audio-frequency signal. One disadvantage of this arrangement is that the limiter does not operate unless the incoming signal has sufficient amplitude to cause the limiting action to take place. When limiting action does not take place, the amplitude variations in the signal result in interfering noise and signal distortion. This means that high-gain RF and IF stages must be used to boost the signal amplitude into the limiter. In a number of FM receivers, two limiters are used in a cascade arrangement to assure that adequate limiting action will take place. Even under these conditions, signals of very low amplitude will result in amplitude modulation into the discriminator, and hence interference and noise.

COMPARING FM DETECTORS

TYPE	FOSTER-SEELEY (DISCRIMINATOR + LIMITER)	RATIO DETECTOR	PHASE-LOCKED LOOP
LIMITATIONS	HIGH GAIN RF AND IF STAGES REQUIRED TO ASSURE LIMITING ACTION	REQUIRES BROAD IF RESPONSE TO ACHIEVE LOW DISTORTION; DIFFICULT TO BALANCE	IMPRACTICAL EXCEPT AS IC PACKAGE
ADVANTAGES	EXCELLENT AUDIO REPRODUCTION; EASY TO BALANCE; AUDIO OUTPUT LEVEL DEPENDS ONLY ON FM DEVIATION	NO LIMITER STAGE REQUIRED; WEAK SIGNALS NOT SUBJECT TO AM INTERFERENCE; AUDIO OUTPUT LEVEL DEPENDS ON BOTH AVERAGE CARRIER LEVEL AND ON FM DEVIATION	EXCELLENT AUDIO REPRODUCTION; EASY TO BALANCE; LOW COST; HIGH RELIABILITY

The important advantage of the ratio detector is that it is not sensitive to short-term amplitude variations in the incoming signal, as was shown in the description of that circuit. Therefore, the ratio detector circuit eliminates the need for the limiter stage, or stages, and does not depend on the use of RF and IF stages of very high gain. The noise level rises with weak signals, but distortion does not take place until the signal amplitude is too low for reception. One disadvantage of the ratio detector is that special care must be taken to balance the two sides of the detector; otherwise some of the insensitivity to amplitude modulation will be lost.

The phase-locked-loop FM demodulator has the outstanding features of reliability, simplicity, and low cost.

De-Emphasis

In your study of FM transmitters, you learned that the signal-to-noise ratio at the high-frequency end of the audio band was improved by passing the modulating signal through a pre-emphasis network. This is possible because there is little signal energy in the high-frequency components, and these can be increased without fear of overmodulation. In this way, noise introduced after pre-emphasis and before de-emphasis will be reduced by the pre-emphasis. The U.S. standards for FM broadcast call for *pre-emphasis* of the higher frequencies in accordance with a 75-μsec time constant. The same *de-emphasis* must be applied at the output of the FM detector to obtain a linear audio output. A series RC network with a 75-μsec time constant serves this purpose. If there is high-frequency falloff in any of the audio stages following the detector, then the *net* rolloff must be in accordance with the equivalent 75-μsec time constant. The de-emphasis circuit is normally placed immediately after the detector since it can also filter IF signals from the audio and prevent the IF signals from feeding through to the audio stages.

DE-EMPHASIS NETWORK

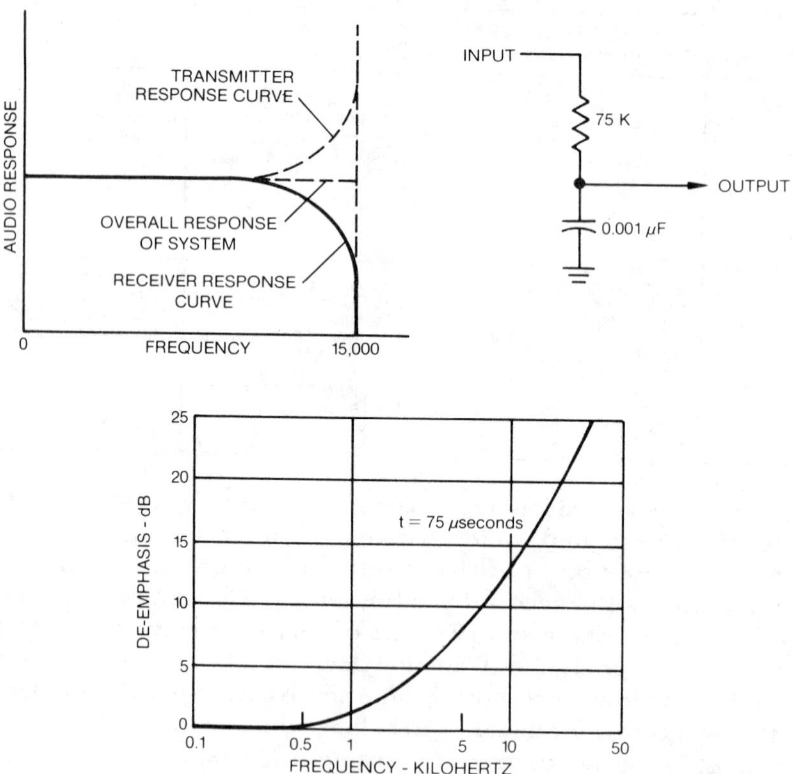

Limiter/Discriminator Circuit

The example of FM limiter and discriminator FM receiver circuit to be discussed here is part of the same receiver previously described with solid-state RF amplifier, oscillator, mixer, and IF stages. The output of the last IF amplifier stage (Q6) is fed to the limiter Q8 and a frequency-discriminator composed of CR19 and CR20. The limiter Q8 is a transistor IF amplifier stage that limits the signal amplitude to a fixed level so that variations in RF signal amplitude are not detected and produced in the audio output. The limiter is needed because the FM detector is of the Foster-Seeley type—that is, amplitude sensitive. The amplitude is limited by operating the stage between cutoff and saturation, which ensures clipping of both positive and negative peaks at all usable signal levels. A low collector voltage on Q8 ensures limiting with low-voltage inputs.

Conversion of the FM signal to an audio signal is accomplished by the Foster-Seeley-type frequency discriminator shown below. The limiter output is coupled from the collector of Q8 to a tuned IF transformer (10.7 MHz). However, this transformer has a center-tapped secondary winding to provide for operation of the Foster-Seeley discriminator. As shown, the quadrature phase shift is obtained from an inductor connected by the center tap, as described previously. Both the primary and secondary windings are tuned to the IF frequency. The audio output is taken from the tap of RC network (K1), and the IF signal leakage is bypassed to ground through the two 220 pF capacitors. The circuit diagram and operation of this discriminator are exactly as described previously. Also, the audio output can be fed to any audio amplifier.

FM RECEIVER LIMITER AND DISCRIMINATOR STAGES

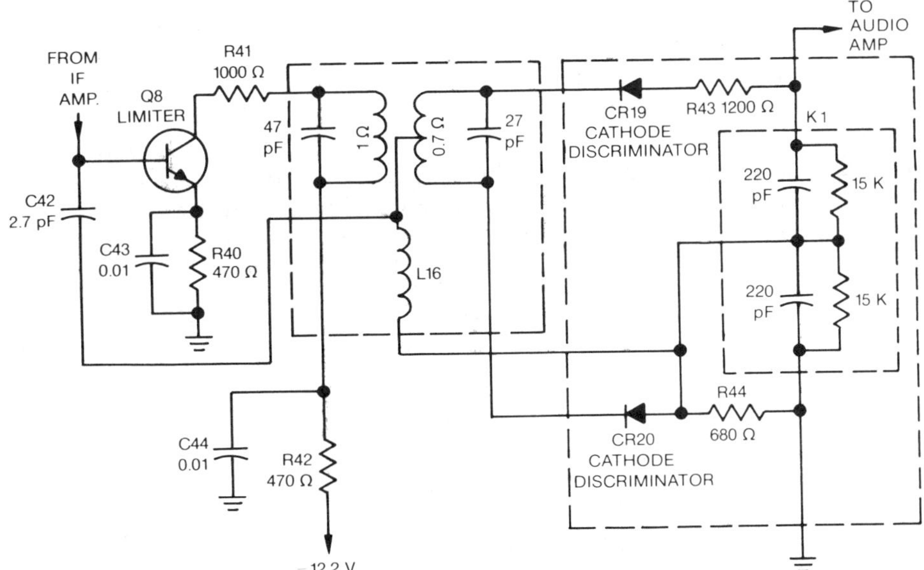

Digitally Tuned FM Receiver

The phase-locked-loop (PLL) system illustrated here in block diagram form provides a method for synchronizing the frequency of an oscillator with the frequency of any suitable reference signal. This system is packaged as an IC and makes it practical and economical to obtain *automatic electronic tuning* in radio and TV receivers. As you will see, it is almost identical with the corresponding part of the digitally tuned CB transmitter that you studied in Volume 3.

The most important part of the phase-locked-loop system is the voltage-controlled oscillator (VCO), which has the capability of being tunable over the entire range of frequencies required by a voltage input. In this respect, it is like the varactor-tuned FM oscillator you learned about when you studied transmitters. For FM automatic tuning, the oscillator that functions as the local oscillator must cover the entire range of LO frequencies (98.8 to 108.6 MHz). The reference-frequency source operates at a lower convenient frequency; consequently, the VCO output is divided down by a counting circuit to the reference frequency or a harmonic of the reference frequency, which is usually produced by a stable crystal oscillator. You will learn how counter circuits work in Volume 5 when you study digital circuits.

A phase detector compares the phase of the VCO signal to the phase of the reference signal and produces a + or − dc error signal when there is a phase (or frequency) difference. For example, if the VCO output were at a higher frequency than the reference frequency, its signal would lead in phase. The phase-comparator output would be a negative tuning voltage. Similarly if the VCO frequency were lower in value than the reference frequency, a positive tuning voltage would be produced.

The tuning voltage is coupled through a low-pass filter that removes rapid variations due to noise, signal, and transient effects. The filter output is used to drive the VCO to a higher or lower frequency as required to keep the phase-detector ouput near zero. The dc tuning voltage is proportional to the phase difference and thus proportional to the frequency difference. When the frequency-divided VCO output frequency exactly matches the reference frequency, no tuning (correction) voltage is produced, and the VCO maintains the correct frequency.

PHASE-LOCKED-LOOP DIGITAL TUNING

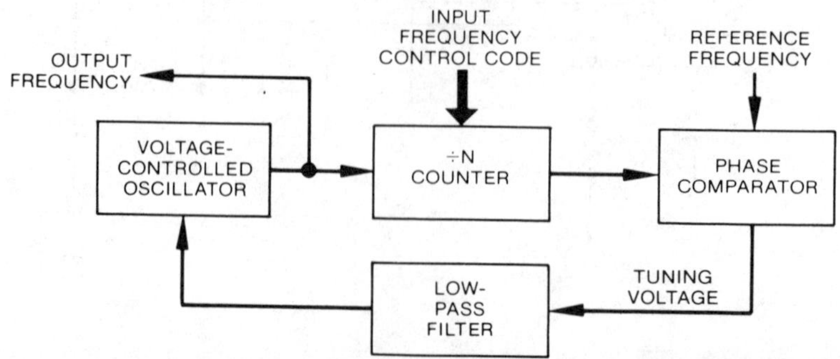

Digitally Tuned FM Receiver (continued)

Using frequency synthesis with a VCO in a phase-locked loop, it is possible to preselect several stations or tune each station in the entire range of stations (100) by properly choosing the proper value for N in the frequency divider. The tuning method consists of simply selecting the number for N in the ÷ N counter and the loop will tune the VCO to the proper local oscillator frequency. In all other respects the FM broadcast receiver, TV, or CB is identical to a more conventionally tuned receiver, except that the tuning is always exact for the channel selected.

A 2.56-MHz crystal-controlled oscillator provides the stable reference signal. This reference signal is divided by 256 (dividing by 2 eight times). Therefore, the crystal-controlled reference signal being fed into the phase comparator is a stable 10-kHz signal. The other signal into the phase comparator is also a 10-kHz signal that is frequency divided down from the voltage-controlled oscillator output signal by N. The VCO is designed to operate as the local oscillator in the 98.8- to 118.6-MHz range, and the oscillator will be tuned above the frequency of the FM station.

PHASE-LOCKED-LOOP AUTOMATIC TUNING

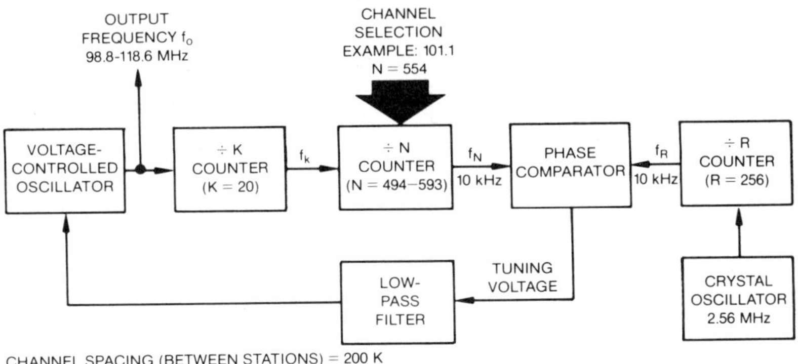

Initially, the VCO output is divided by 20 by a fixed counter. Thus, the frequency range of signals representing the VCO frequency is now 4.94 to 5.93 MHz. To select a particular station or channel, a number is now programmed into the variable (channel-select) counter. For the FM band and the system shown, this number is any number (N) from 494 to 593. A value of N equal to 494 produces a 98.8-MHz VCO output for the first station on the dial (88.1 MHz); a value of N equal to 593 produces a 118.6-MHz VCO output for the last station on the dial (107.9 MHz).

For example, to tune to a station at 100.1 MHz (100 on the dial), the value of the number N would be calculated as follows: 100.1 + 10.7 MHz = 110.8 MHz: 110.8/20 MHz = 5.54 MHz and N = 554; 5.54MHz/554 = 10 kHz.

When the selected value of N is programmed into the ÷ N counter, the counter output will be exactly 10 kHz if the VCO is on frequency. Comparison of the divided VCO frequency with the reference signal at exactly 10 kHz produces a tuning error voltage to drive the VCO to the proper frequency and hold it exactly at that frequency.

FM Stereo

Most high-fidelity home audio systems use stereophonic reproduction, and stereo recording—both tape and disk—has almost entirely replaced monaural recording. You learned about stereo audio systems in Volume 2 and how the two channels required for stereo (left and right) are prepared for transmission in Volume 3. The principle of FM stereo transmission will be reviewed here.

The format for stereophonic transmission is set by the FCC, and requires the use of a *main* channel and a *subchannel*. Furthermore, the information transmitted on the main channel must be transmitted so that it can be received, without degradation, by a monophonic FM receiver. Monaural or monophonic reproduction of a stereo signal is done by simply adding the left and right outputs to form a single channel (L + R).

FORMATION OF FM STEREO SIGNAL

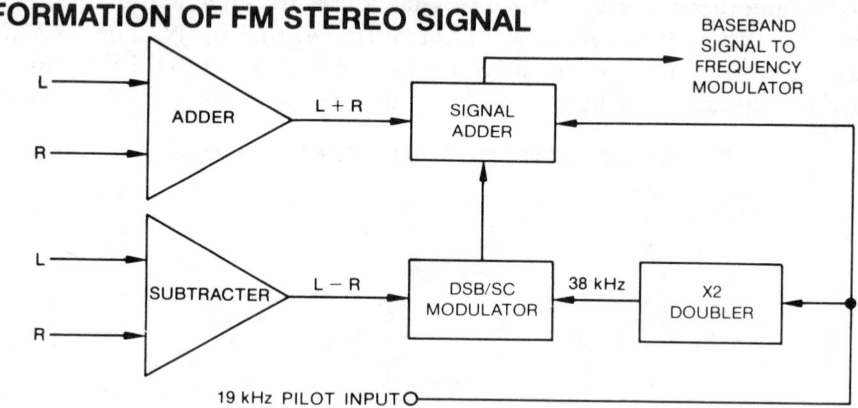

STEREOPHONIC BASEBAND SPECTRUM

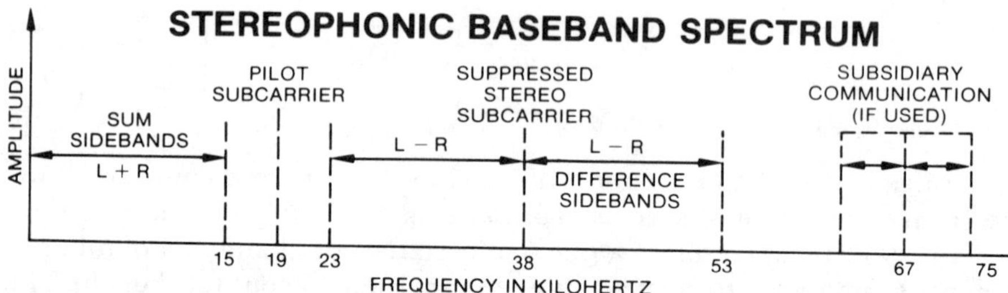

At the transmitting station, the L (left) and R (right) signals are combined to form the sum (L + R) and difference (L − R) signals. The sum signal represents the total monophonic information. The L − R signal is used to amplitude modulate a 38-kHz subcarrier in a double-sideband (DSB) suppressed-carrier modulator. The 38-kHz signal is derived from a 19-kHz ± 2-Hz *pilot* subcarrier. The L + R signal (50 Hz–15 kHz), the (L − R) sideband signal (23 kHz–53 kHz), and the 19-kHz pilot are added together and used to frequency modulate the carrier. The composite baseband stereo signal spectrum and a block diagram showing the formation of the FM stereo signal are shown in the figure above. An extra channel for subsidiary communciation, also sometimes added to the modulation, will be described later.

FM Stereo Matrix Decoder

Receiving stereo programs requires recovery of both the L + R and L − R signals and recombining them to produce the original left and right channels of information. The output from the detector of an FM broadcast receiver contains the entire baseband spectrum from 50 Hz to 53 kHz. The sum channel (L + R) is readily recovered by passing this signal through a 15-kHz lowpass filter. Recovery of the L − R signal requires the *reinsertion* of the 38-kHz suppressed carrier in the *correct phase relative to the sidebands*, and the envelope detection of the 38-kHz amplitude-modulated signal. The 38-kHz signal is obtained by using the 19-kHz pilot signal to phase lock the locally generated 38-kHz source or by doubling the 19-kHz pilot signal recovered from the discriminator output by a narrow-band filter.

The *FM stereo matrix decoder* is so named because it uses a *resistive matrix* to recombine the L + R and L − R signals to recover the original left and right signals. This is the form of most of the early FM stereo decoders, but it has been largely replaced by phase-locked-loop circuits.

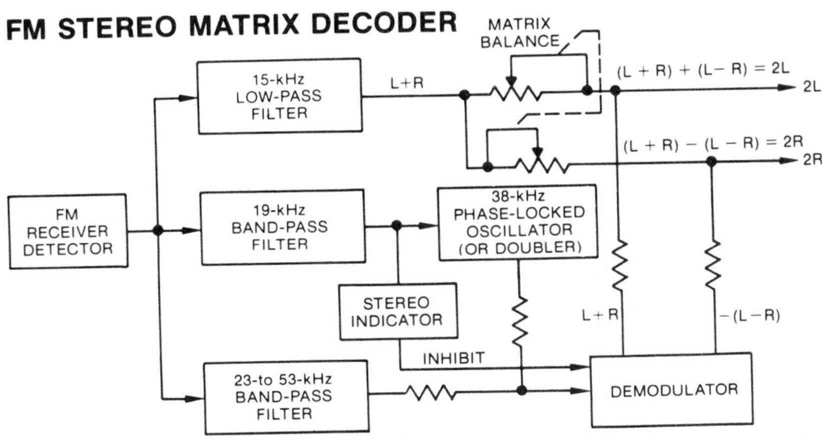

The composite baseband signal (containing the L + R, pilot, and DSB L − R signals) is fed to three tuned filters as shown. The L + R signal is recovered by the 15-kHz low-pass filter. The output of the 19-kHz pilot filter is fed to a 38-kHz oscillator or frequency doubler to generate a 38-kHz reference signal that demodulates the L − R signal.

The output of the 23- to 53-kHz filter (the upper and lower sidebands of the stereo subcarrier) is added to the 38-kHz reconstituted carrier and demodulated (detected) to provide the original L − R signal. The L + R and L − R signals are added algebraically, as shown, to produce the left- and right-channel signals.

In most FM receivers, the presence of the 19-kHz pilot signal is detected, and the detected signal is used to light a stereo-indicator light. When no pilot signal is present (indicating monophonic transmission), the L − R channel is inhibited and the L + R signal is supplied to both audio channels of the receiver automatically.

FM Stereo Time-Multiplex Adapter/Decoder

The time-multiplex adapter/decoder is an envelope-detection method wherein the composite L + R and L − R stereo signals are sampled at the phase-synchronous rate of 38 kHz. This approach to decoding has an advantage over the conventional matrix method in that neither a 15-kHz low-pass or a 23- to 53-kHz bandpass filter is required, and the balance problem is eliminated. This can result in significant cost reduction and improved system performance.

TIME-MULTIPLEX ADAPTER/DECODER

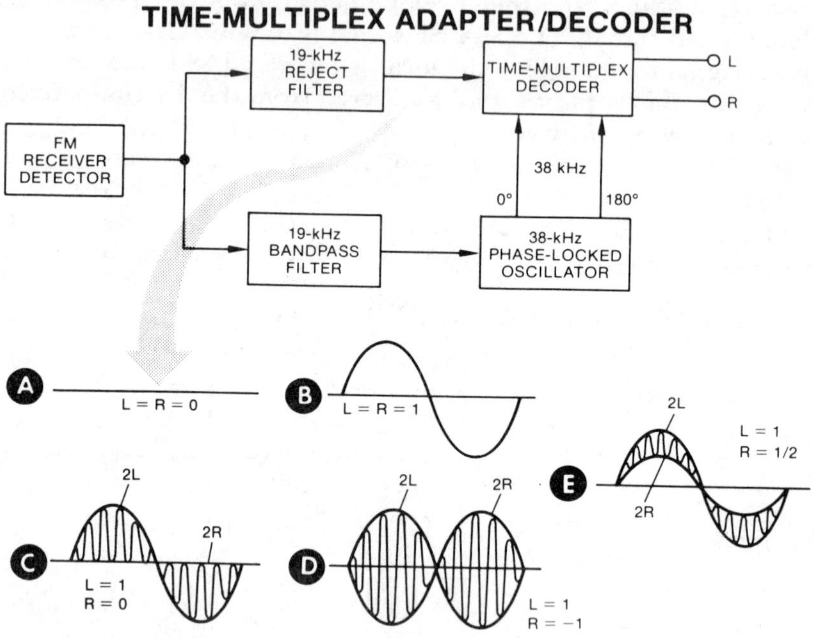

WAVEFORMS AT FM RECEIVER DETECTOR OUTPUT WITH PILOT SIGNAL REMOVED

Consider first the composite signal that exists at the input to the decoder for various combinations of L and R signals. For convenience, the 19-kHz pilot signal is not shown. If L and R signals are both zero, than L + R and L − R are also zero (A). Since the subcarrier is suppressed, the detector output (after the 19-kHz filter) is also zero (A). If L and R are equal and in phase (B), L − R equals zero, and sampling will yield identical outputs for both L and R. When L and R are not identical (C through E), the composite signal contains a 38-kHz component that has *envelopes* of 2L and 2R. In the time-multiplex system, the composite signal is sampled at a phase-coherent 38-kHz rate, and the envelope signals are thereby recovered—yielding the L and R channel outputs directly. The time-multiplex decoder or sampler samples the positive peaks at the 0-degree point (in phase) and the negative peaks at 180 degrees (out of phase) at the 38-kHz reference rate. As you can see from the illustration, this directly results in separation of the left and right audio signals at the output of the time-multiplex decoder without the need of the filters used in the matrix decoder described previously.

FM Stereo Time-Multiplex Adapter/Decoder (continued)

The operation of an FM stereo time-multiplex adapter/decoder can be illustrated by referring to the transformer-driver, diode-detector configuration shown. The composite multiplex signal is added to the in-phase (0-degree) 38-kHz reference in transformer T1 and fed to diodes CR1 and CR2. Similarly, the composite signal is added to the out-of-phase (180 degree) 38-kHz reference signal and fed to diodes CR3 and CR4. These two composite waveforms are shown in diagrams A and B.

The composite signals at points A and B are rectified by diodes CR1–CR2 and CR3–CR4, respectively, and produce the output signals shown by waveforms 1 through 4. The outputs from diodes 1 and 4 (waveforms 2 and 3) are summed to produce the left-channel output, and the outputs from diodes 2 and 3 (waveforms 1 and 4) are combined to provide the right-channel output. The two 150-K resistors (effectively in parallel) and the 0.001-μF capacitor constitute the 75-μsec de-emphasis network (75 K × 0.001 = 75 μsec).

TIME-MULTIPLEX ADAPTER/DECODER

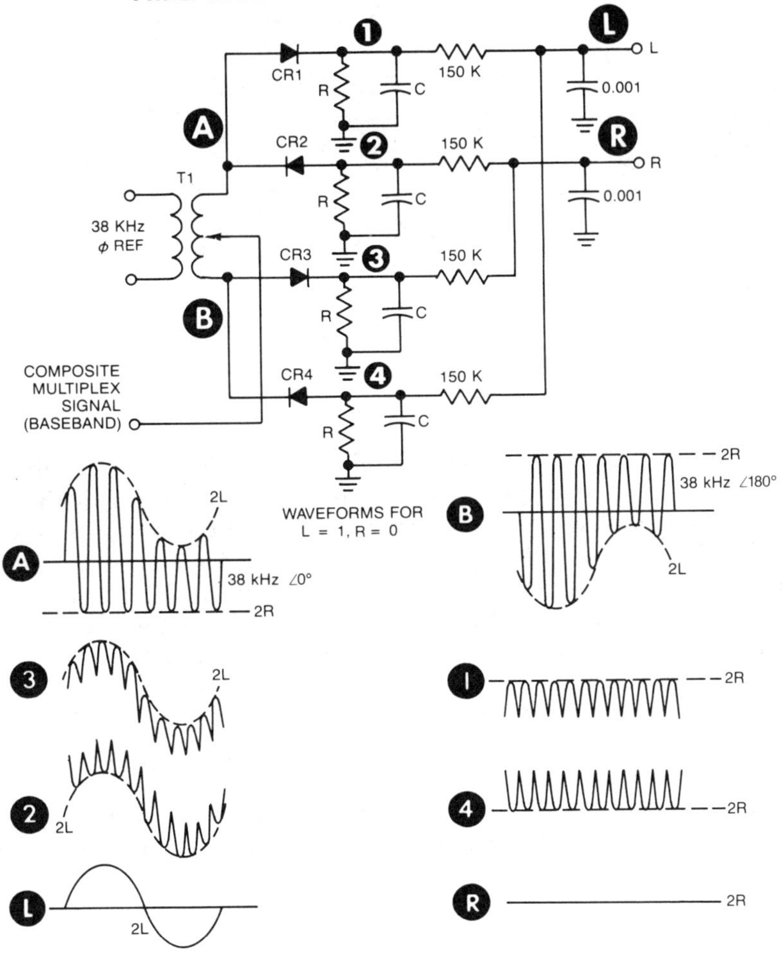

Modern IC FM Stereo Demodulators

The improvements in phase-locked-loop technology led to the development of IC stereo multiplex decoders that do not require the use of *any* resonant circuits, provide automatic stereo/monaural switching, and include a driver to operate the stereo-lamp indicator. Thus, the entire stereo decoding is done in a single IC. The heart of the IC stereo demodulator is a 76-kHz VCO whose output is counted down by three frequency dividers to provide a 38-kHz signal and two 19-kHz signals in phase quadrature.

MODERN IC FM STEREO DEMODULATOR

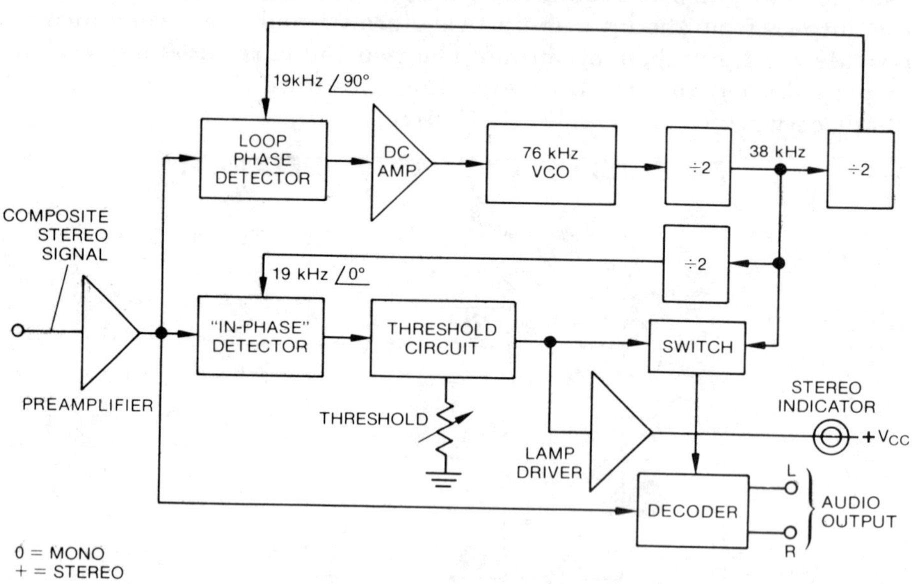

The composite stereo signal containing the 19-kHz pilot signal is compared to the 19-kHz quadrature signal (at 90 degrees) in the loop phase detector, and the resultant error signal is used to phase lock the 76-kHz VCO (RC-controlled oscillator). As you remember from your study of phase-locked loops in Volume 3, the output from a VCO is at 90 degrees with respect to the input signal. Shifting the phase of the feedback signal by 90 degrees produces an in-phase output from the VCO needed to properly decode the FM stereo signal. A second phase detector compares the 19-kHz in-phase signal with the pilot signal. If the pilot signal is present it will, by nature of the 0-degree phase relation, produce a unipolar dc output. If this output exceeds an internal threshold, it indicates that a stereo program is present and activates the indicator lamp and the switch that allows the 38-kHz reference into the decoder.

Modern IC FM Stereo Demodulators (continued)/Subsidiary Communication Authorization (SCA)

The phase-locked 38-kHz signal, switched into the decoder when a stereo program is present, is combined with the composite stereo input to provide the left- and right-channel signals, as described previously. The decoder can be either the matrix-decoder or time-multiplex configuration. De-emphasis for the L and R channels is normally applied external to the IC demodulator. Modern stereo FM receivers almost invariably use a single IC for FM stereo demodulation, although not all are specifically of the type described on the previous page. Some still require externally tuned components for operation.

Now let us move on to our next subject—the study of subsidiary communication authorization (SCA).

FM broadcast stations are often authorized to transmit a second set or more of DSB signals to carry special information. For example, special commercial messages, directed to specific places, or tones to cut out advertising can be transmitted via SCA (subsidiary-communication authorization) channels. These channels have a bandwidth of 10 kHz (±5 kHz) and are formed just like the L − R channel except that the carrier frequency is higher—for example, 67 kHz. This frequency is chosen so that the SCA channel falls between the L − R channel upper-sideband limit (53 kHz) and the fourth harmonic (76 kHz) of the 19-kHz pilot signal, as shown in the diagram. As you can see, the SCA channel is only another sideband signal on the carrier which can be recovered by the filtering and demodulation techniques described earlier.

The SCA channel can have subcarriers on it which provide up to five separate information signals—for example, tones of various frequencies to allow for decoding at the receiver using appropriate filters. Thus, either voice transmissions and/or tones can be used to provide control information for use at the receiver. In many cases, music in public places like airports, restaurants, shopping malls, and department stores is obtained from FM stereo receivers. The SCA channel is used to provide signals that inhibit the audio output during commercials or news broadcasts so that only the music is heard.

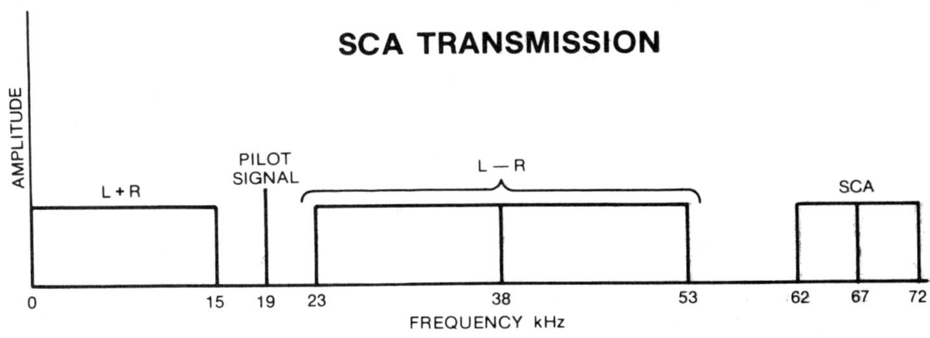

Automatic Frequency Control (AFC)

Because FM receivers operate in the VHF/UHF range, it is difficult to design the local oscillator so that it will be stable under different temperatures and line-voltage conditions, etc. Thus, many FM receivers are equipped with automatic frequency control (AFC). Of course, modern receivers using a frequency synthesizer do not need this feature because the local oscillator signal is stable and accurate. The AFC signal is derived from the discriminator output and is fed to the LO, which has an additional voltage-controllable tuning element such as a varactor to provide for limited tuning range (usually a few hundred kilohertz).

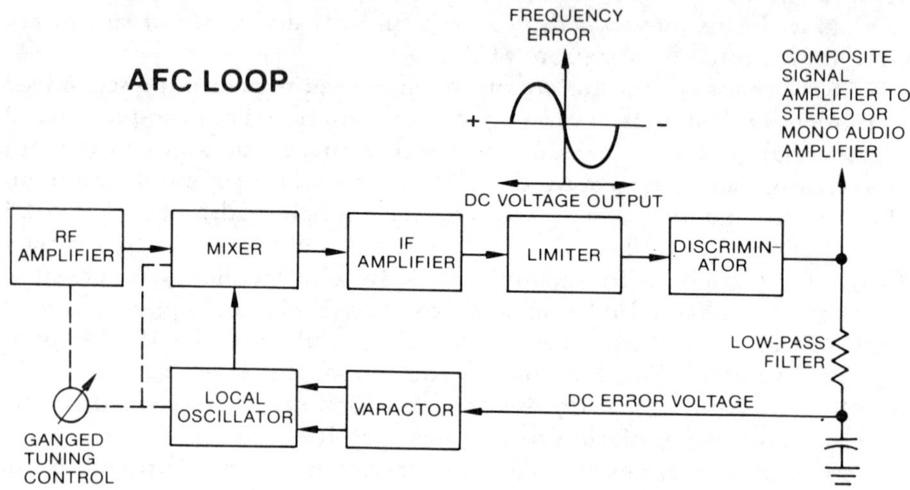

As you will recall, the output from some FM discriminators, such as the Foster-Seeley discriminator, is zero at the carrier intermediate frequency. With no modulation, if the carrier is mistuned, a dc offset (error) voltage is produced, and is proportional to the mistuning. This voltage can be used in a feedback loop to correct the tuning error, as shown in the diagram. If the receiver is not quite properly tuned or the LO drifts slightly, a dc voltage appears at the discriminator output in addition to the audio signals. A low-pass filter (typically, with a time constant of 0.1 to 0.5 second) filters out the audio signal, leaving only the dc frequency error component. If this is applied in the proper polarity to a voltage-controllable tuning element such as a varactor in the LO tuned circuit, the LO will be retuned until the dc voltage drops to zero (or very close to it), which means that the LO is properly tuned. Thus, the LO is kept properly tuned for best reception. Since the ratio detector does not have zero output at the correct tuning point, it cannot be used in a simple AFC loop as shown above. Receivers with an AFC loop usually have a switch to disable it. This is because it is sometimes difficult to listen to a weak station on a channel adjacent to a strong station because the AFC will pull the receiver tuning to the strong station.

Review of FM Receivers

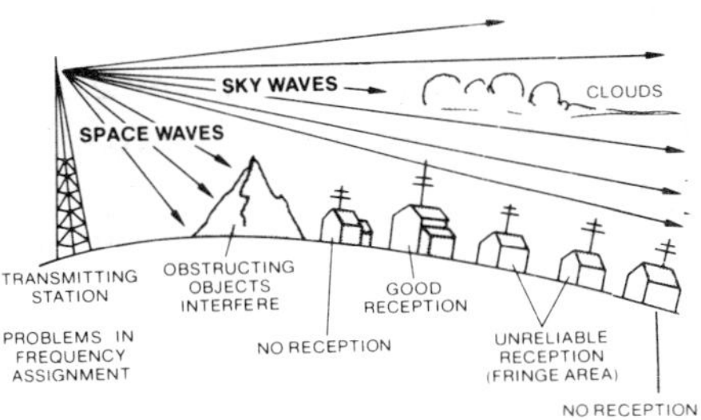

1. **FREQUENCY MODULATION (FM)** is used where high fidelity and noise immunity are important. For full benefit, it requires large bandwidths and is broadcast on VHF, where transmission is largely limited by *line of sight*.

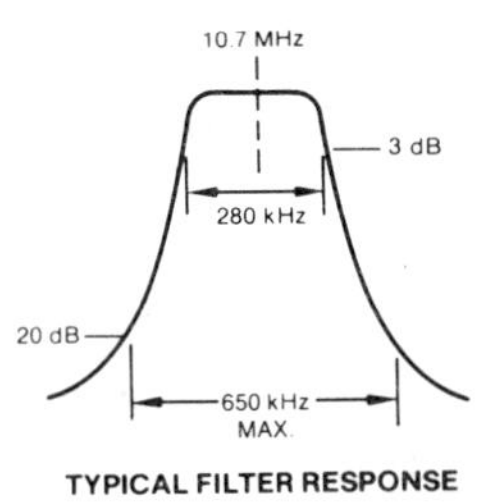

2. **FM RF, LO AND MIXER STAGES**, except for frequency, are like those in other superhets. The *IF amplifier* must be carefully designed for the right *selectivity* curve.

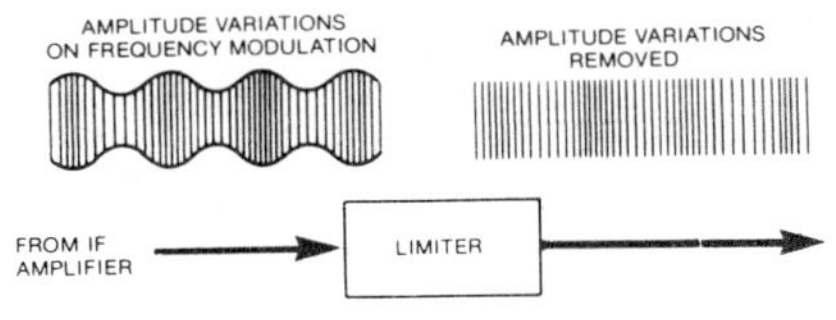

3. **FM DETECTORS** must produce a voltage that changes linearly with carrier frequency. Most popular are the *Foster-Seeley*, *ratio detector*, and *phase-locked loop* types. Some are sensitive to AM and must be preceded by a *limiter*.

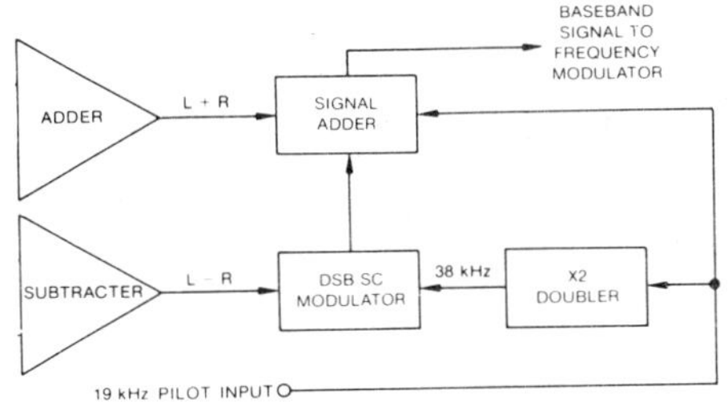

4. **STEREO REPRODUCTION** requires separate left and right channels. Transmission uses a *main channel* and a *subchannel*. Reception requires a *stereo multiplex demodulator*.

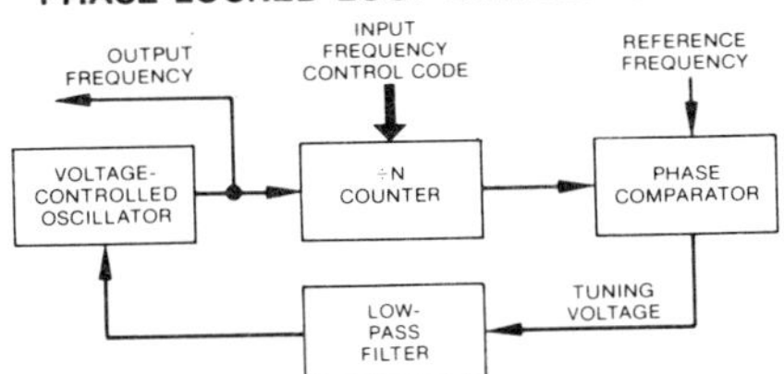

5. **AUTOMATIC FREQUENCY CONTROL (AFC)** uses an error signal from the discriminator to keep the LO on frequency. Some receivers use *PLL synthesizers*, which provide their own AFC.

Self-Test—Review Questions

1. Describe the basic difference between frequency and amplitude modulation.
2. Why is FM used?
3. What are the basic differences between an FM and an AM receiver?
4. Describe the operation of FM detectors. Why is limiting required? Discuss limiters. Compare various FM detectors.
5. Why is de-emphasis used? How does it function?
6. Describe how FM stereo is received and decoded.
7. What is SCA?
8. What is AFC? How does it work? Why is it needed?
9. Describe briefly how a digitally tuned FM receiver works. Do you need AFC on a digitally tuned FM receiver?
10. Why is FM confined to higher frequencies? Are FM broadcast antennas different from AM broadcast antennas?

Learning Objectives—Next Section

Overview—In the section that follows you will study communication receivers, including those for CB, fleet, amateur, and satellite communication.

TYPICAL FLEET COMMUNICATION FM RECEIVER

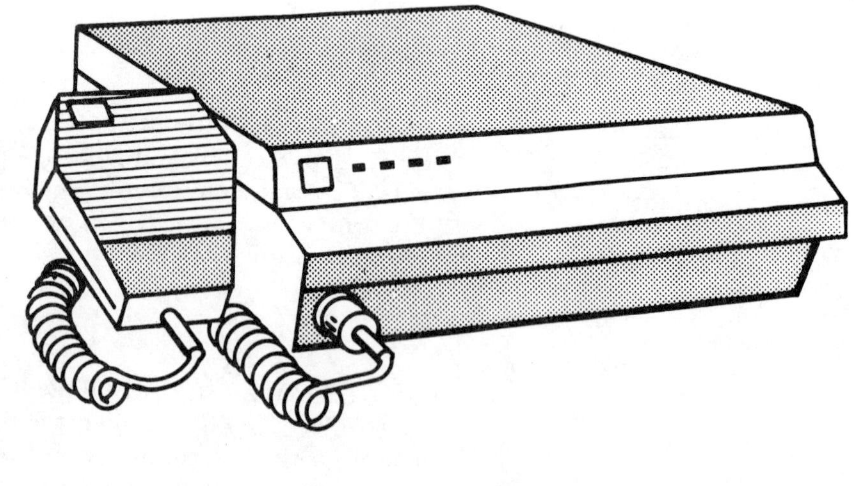

Introduction

This section will be devoted to a discussion of communication receivers. These receivers constitute the largest group of receivers except for the standard broadcast AM and FM radio receivers. In most cases, the transmitter and receiver are combined into a single transceiver to be used for *both* transmission and reception. This is done because costs can be reduced by using common circuits for both applications. The major categories and frequency assignments of public and private communication systems are shown in the table on the following page. Since, as indicated by its name, a communications receiver is primarily for reception of information, intelligibility is more important than fidelity and tone quality. Also, it must often operate at reduced input signal levels (compared to those in broadcast reception) and cope with high levels of interference.

The more than 10 million CB operators in the United States represent the largest group of communication receiver users. In order to minimize interference, CB transmitter power is limited to 4 watts, and there are 40 different frequencies of operation identified as channels 1 through 40. The CB frequency band is at approximately 27 MHz, with 10-kHz separation between channels.

Fleet communication systems represent the next largest group of operators. These systems are reserved for commercial trucking, taxi, and other basic service organizations, and their power output can be up to 50 watts to provide greater distance coverage. In this service, cochannel and adjacent channel interference is normally less severe than for amateur and CB operation, because frequency assignments are more restrictive.

Other large users of radio communication are public systems such as police, fire, medical emergency, and other community-service groups. These are mostly limited power systems, because their vehicle operations are usually only short distances from the home stations.

Many radio amateurs (over 250,000) use special bands for communications, as shown on the following page. Amateur radio operation includes almost all forms of communication, from telegraphy to TV. Amateur equipment includes some of the most sophisticated receiving systems, since much operation is at minimum signal level and at high density of frequency-band population.

Another special category includes the microwave relay stations used by the telephone, TV, and data-processing networks. Such systems also use satellite relay to speed messages around the world.

The receivers to be considered on the pages that follow operate on the same principles as those described previously. Only brief descriptions will be given of the individual stages, with the emphasis on special features and the interconnection of the stages to make up a complete system.

COMMUNICATION RECEIVERS

FREQUENCY CHART

Amateur

1,800 -2,000 kHz
3.500- 4.000 MHz
7.000- 7.300
14.00 - 14.35
21.00 - 21.45
28.00 - 29.70
50.00 - 54.00
144.0 - 148.0
220.0 - 225.0
420.0 - 450.0
1,215 -1,300
2,300 -2,450
3.300- 3.500 GHz
5.650- 5.925
10.00 - 10.50
24.00 - 24.25
48.00 - 50.00
71.00 - 84.00
152.0 - 170.0
200.0 - 220.00
240.0 - 250.00
Above 275.0

Citizens Radio (Personal radio services)

26.96 - 27.23 MHz
462.5375-462.7375
467.5375-467.7375

Land Mobile (Communication on land between base stations and mobile stations or between mobile stations)

Land Transportation (Taxis, trucks, buses, railroads)

30.56 - 32.00 MHz
33.00 - 34.00
43.68 - 44.61
150.8 - 150.98
152.255 - 152.465
157.45 - 157.725
159.48 - 161.575
451.0 - 454.0
456.0 - 459.0
460.0 - 462.5375
462.7375- 467.5375
467.7375- 512.0
1,427 -1,435

Broadcast Remote Pickup

1,605 -1,715 kHz
26.10 - 26.48 MHz
161.625 - 161.775
166.25
170.15
450.0 - 451.0
455.0 - 456.0

Land Mobile (Communication on land between base stations and mobile stations or between mobile stations)

Public Safety (Police, fire, highway, forestry, and emergency services)

1,605 -1,750 kHz
2,107 -2,170
2,194 -2,495
2,505 -2,850
3.155 - 3.400 MHz
30.56 - 32.00
33.01 - 33.11
33.41 - 34.00
35.19 - 35.69
37.01 - 37.43
37.89 - 38.00
39.00 - 40.00
42.00 - 42.95
43.19 - 43.69
44.61 - 46.60
47.00 - 47.69
150.98 - 151.4825
153.7325- 154.46
154.6375- 156.25
158.715 - 159.48
162.0125- 173.2
451.0 - 454.0
456.0 - 459.0
460.0 - 462.5375
462.7375- 467.5375
467.7375- 512.0
1,427 -1,435

Industrial (Power, petroleum, pipeline, forest products, factories, builders, ranchers, motion picture, press relay, etc.)

1,605 -1,750 kHz
2,107 -2,170
2,194 -2,495
2,505 -2,850
3.155 - 3.400 MHz
4.438 - 4.650
25.01 - 25.33
27.28 - 27.54
29.70 - 29.80
30.56 - 32.00
33.11 - 33.41
35.00 - 35.19
35.69 - 36.00
37.00 - 37.01
37.43 - 37.89
42.95 - 43.19
47.43 - 49.60
151.4975- 152.0
152.465 - 152.495
152.855 - 153.7325
154.46 - 154.6375
157.725 - 157.755
158.115 - 158.475
173.2 - 173.4
216.0 - 220.0
451.0 - 454.0
456.0 - 459.0
460.0 - 462.5375
462.7375- 465.5375
467.7375- 512.0

Citizens Band (CB) Radio

The purpose of the CB radio is to provide citizens with a means of communicating meaningful information such as highway and weather conditions while traveling, reporting breakdowns or sickness, and allowing the traveler to communicate with the outside world. CB is not restricted to mobile uses, and it can be used between fixed stations as well.

There are over 10 million CB sets licensed for use in the United States today. Most are transceivers (transmitter/receiver combinations). Manufacturers of CB transceivers provide information to the buyer of the restrictions and licensing requirements for each transceiver. Actually, it is only the transmitter portion that requires licensing. Most manufacturers supply a license application form with the transceiver. No special skill is required for CB operation, but a knowledge of rules and regulations is required. CB operates at about 27 MHz, and CB transceivers use AM for information transfer.

With a mobile installation and an average antenna, the signal doesn't usually travel more than 5–10 miles (8–16 km) and often much less. Currently, there are 40 channels available in the United States, and maximum output power of 4 watts for either fixed or mobile stations.

U.S. Citizens Band Frequencies (40-Channel)

CHANNEL	FREQUENCY (MHz)	CHANNEL	FREQUENCY (MHz)
1	26.965	21	27.215
2	26.975	22	27.225
3	26.985	24**	27.235
4	27.005	25**	27.245
5	27.015	23**	27.255
6	27.025	26	27.265
7	27.035	27	27.275
8	27.055	28	27.285
9	27.065	29	27.295
10	27.075	30	27.305
11	27.085	31	27.315
12	27.105	32	27.325
13	27.115	33	27.335
14	27.125	34	27.345
15	27.135	35	27.355
16	27.155	36	27.365
17	27.165	37	27.375
18	27.175	38	27.385
19	27.185	39	27.395
20	27.205	40	27.405

**NOTE VARIATION IN SEQUENCE.

The FCC, which is the U.S. regulating agency, has some sophisticated equipment for monitoring CB operation, and their prime concern is to encourage operators to follow the rules so that everyone can use CB for the purpose it was designed—citizens' communications.

Citizen Band (CB) Radio (continued)

Prior to expansion of the band to 40 channels in 1977, CB receivers and transmitters were limited to 23 channels. These ranged in frequency from 26.965 to 27.255 MHz, or a total spread of less than 300 kHz. Currently, 40 channels are available covering the bands from 26.965 MHz to 27.405 MHz in 10-kHz steps, as shown on the previous page. The receiver to be described here is the receiver section of the 40-channel CB transceiver that you studied in Volume 3. There, you learned how the frequency synthesizer operated to provide for 40-channel operation with just three crystals. You also learned that the IF was 4.3 MHz, and that the synthesizer basically produced the LO signal; the transmit signal was generated by mixing this LO signal with the output from a 4.3-MHz crystal oscillator. The sum signal from this mixer (LO + 4.3 MHz) provides the transmit signal.

SIMPLIFIED BLOCK DIAGRAM OF RECEIVER IN CB TRANSCEIVER

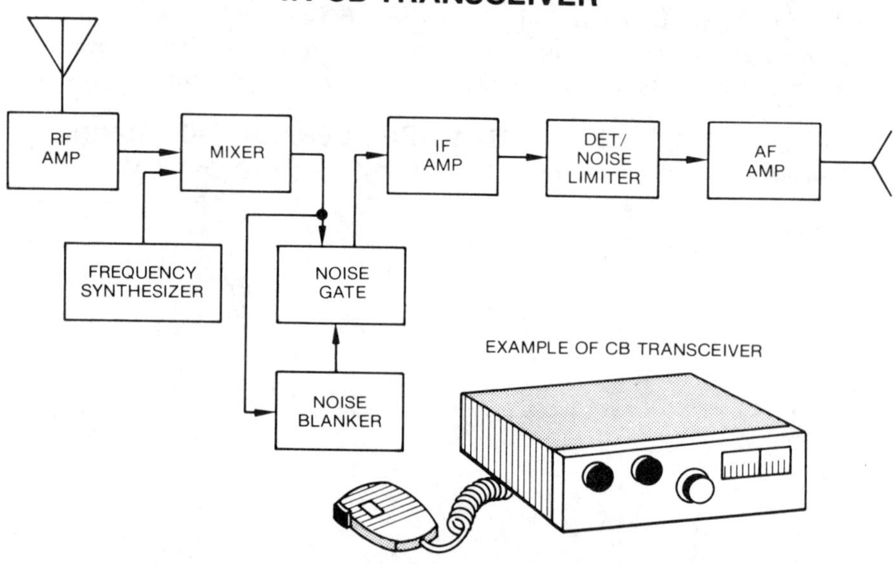

This receiver contains signal-processing techniques and special features widely employed in modern AM communications receivers and is, therefore, a good medium for introducing this subject. A feature of the receiver is the synthesizer that allows for operating on any of the 40 channels by merely setting a selector switch for the desired channel.

Other features are a noise-blanking circuit to eliminate noise when no signal is being received, audio noise limiting, and the use of crystal filters in the IF section to provide sharply tuned IF amplification (which prevents interference from adjacent channels). Otherwise, the circuits and their arrangement are essentially the same as those described for a conventional AM receiver.

CB Radio RF Amplifier and Mixer

Input signals to the receiver are obtained when the transmitter is not keyed on. In the receive mode, the antenna is automatically connected to the receive input circuit. The antenna signal is coupled through the tuned transformer T1 to the base of RF amplifier Q1. Automatic gain control is provided from the Q21 collector to the emitter of the RF amplifier to compensate for input-signal level variations. The ouput of the RF amplifier is tuned-transformer coupled to the base of the receiver mixer Q2. At the same time, the proper synthesized local oscillator signal is also coupled from the cathode of CR201 to the base of Q2. Automatic gain control is also provided to the mixer from the Q22 collector. Thus, the mixer receives the incoming signal, and the switching of the local oscillator determines the channel selection. The output of the mixer Q2 is applied to both T11 of the noise-blanking circuit and to IF transformer T3. The IF transformer is tuned to 4.3 MHz, which is the difference between the channel operating frequency and the output frequency of the synthesizer.

CB RECEIVER RF AMPLIFIER AND MIXER STAGES

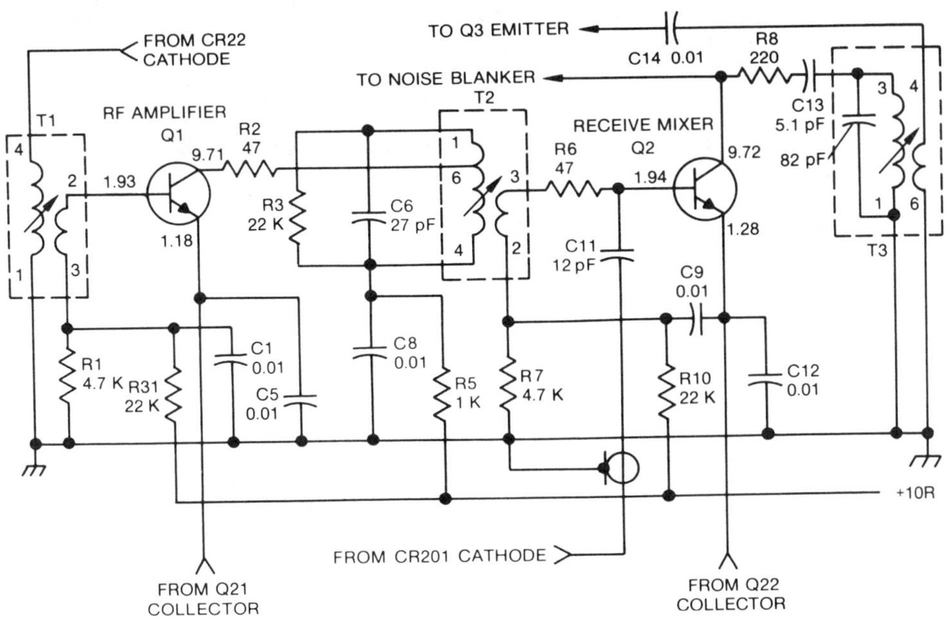

While the tuned circuits in the front end of the CB receive portion are adjustable, they are not adjusted as the channel is changed. Instead, the RF tuned circuits are made sufficiently broadband so that no tuning is required over the range of input frequencies (approximately 27 to 27.5 kHz). Thus, it is the setting of the LO frequency in conjunction with the high selectivity of the IF circuits that provides the frequency selection and selectivity of the receiver.

CB Radio Noise Blanker

The noise-blanking circuit in a communication receiver is used when receiving weak signals in the presence of noise pulses or impulse noise such as automobile ignition interference. It effectively gates the receiver off during the period of the noise pulse. If the impulses are short, the gated intervals are also short, and the impulse noise is deleted without loss of intelligibility. As with most noise blankers, the system described here uses a separate relatively broadband IF amplifier to operate the noise-gate generator, which in turn gates the IF amplifier input. The noise-gate IF is relatively wide-band to allow the circuit to react before the narrow-band IF amplifier. The separate IF is necessary also to be able to detect the duration of the noise pulse to establish the gate time in the signal IF. When the blanker is switched out, the signal from the mixer is coupled directly to the first IF stage. Setting the switch to the noise-blanking position couples the signal from the mixer through two noise-blanker IF stages, a noise detector, pulse amplifier, and noise gate.

Noise-blanking circuits are only useful when the noise is *impulsive*—that is, when it consists of sharp spikes of noise like ignition noise or electrical equipment noise. It is *not* useful against the background hiss noise generated by the receiver. The block diagram on this page shows the basic elements in the noise blanker.

BLOCK DIAGRAM OF NOISE BLANKER CIRCUIT

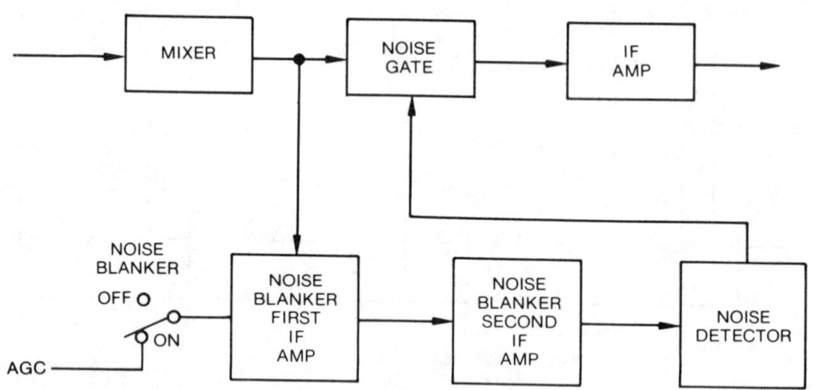

The mixer output is fed to a noise gate that is between the mixer and the first signal IF. The mixer also feeds the signal to the noise-blanker IF amplifier, which in turn feeds a second noise IF amplifier. These differ from the signal IF in having a wider bandwidth. The first noise-blanker IF is decreased by the AGC voltage when strong signals are present, since noise blanking is less necessary in this case. The IF output from the second noise IF amplifier is fed to the noise detector, which detects pulses exceeding the threshold. The gate pulses generated are fed to the noise gate to blank the IF input to the signal IF amplifier.

CB Radio Noise Blanker (continued)

With the noise-blanker circuit switched in, AGC voltage turns on the first noise-blanker IF amplifier Q10. The mixer output is coupled through the tuned IF transformer T11 and amplified in two noise IF stages, Q10 and Q11. The amplified signal is detected by the noise detector CR9. CR9 conducts only on the positive portions of the IF signal, charging C62 to a positive voltage that corresponds to the peak amplitude of the noise pulses. The audio portion of the signal is removed by the audio filter consisting of R54, C62, C63, and R57. The detected positive voltage forward-biases the pulse amplifier Q12. When Q12 conducts, the collector voltage decreases, and this negative-going pulse is coupled through pulse-shaper network C65, L5, and C15 to base of noise gate Q3.

REARRANGEMENT OF NOISE BLANKER CIRCUIT

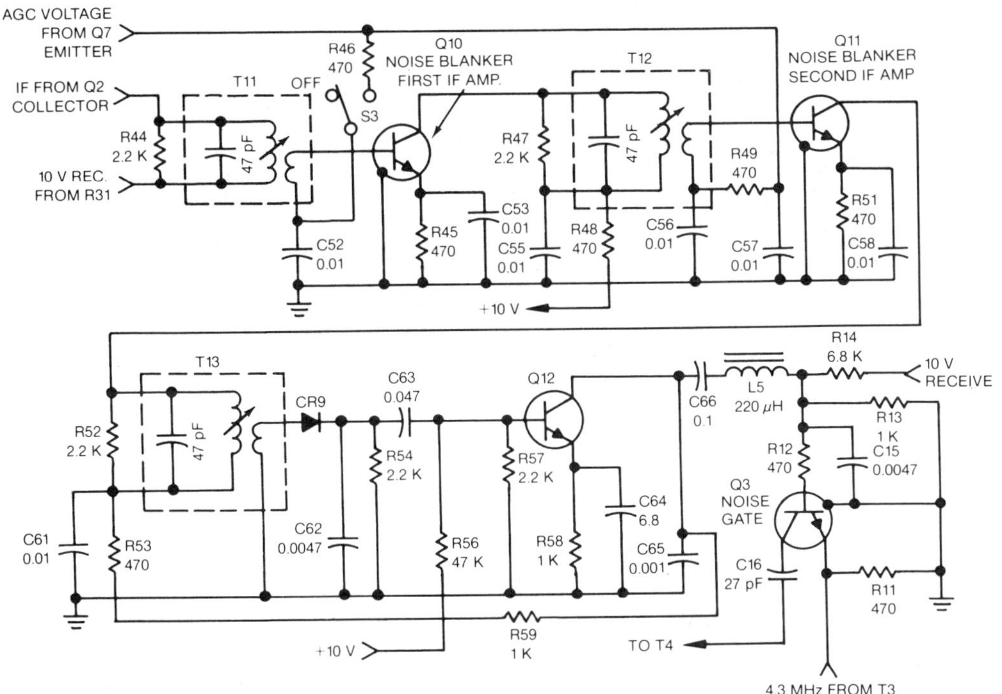

A negative-going voltage on the base of Q3 acts as a reverse bias to cause Q3 to cut off. Since the negative bias on Q3 corresponds to IF noise pulses present on the Q3 emitter, the noise gate is cut off only when the noise pulse or pulses are present.

The AGC voltage is connected to Q10 and Q11, the noise-blanker first and second IF stages, to vary the gain of the noise-blanker amplifiers. The negative AGC voltage acts as a reverse bias to decrease the gating-pulse amplitude when a large amplitude signal is received. Such a large amplitude signal overrides the noise, and the noise-blanker circuitry is not necessary while receiving high-level signals.

CB Receiver IF Amplifiers

Bandwidth and selectivity of the receiver are controlled by two 4.5-MHz crystal filters and the three IF amplifiers that have coupling transformers also tuned to 4.5-MHz.

When the noise-blanking circuit is not gating the signal off, the mixer output is coupled through the noise gate Q3 to the crystal input filter through matching transformer T4. The output of the filter is also impedance-matched to the first IF amplifier by tuned transformer T5.

CB RECEIVER IF AMPLIFIER

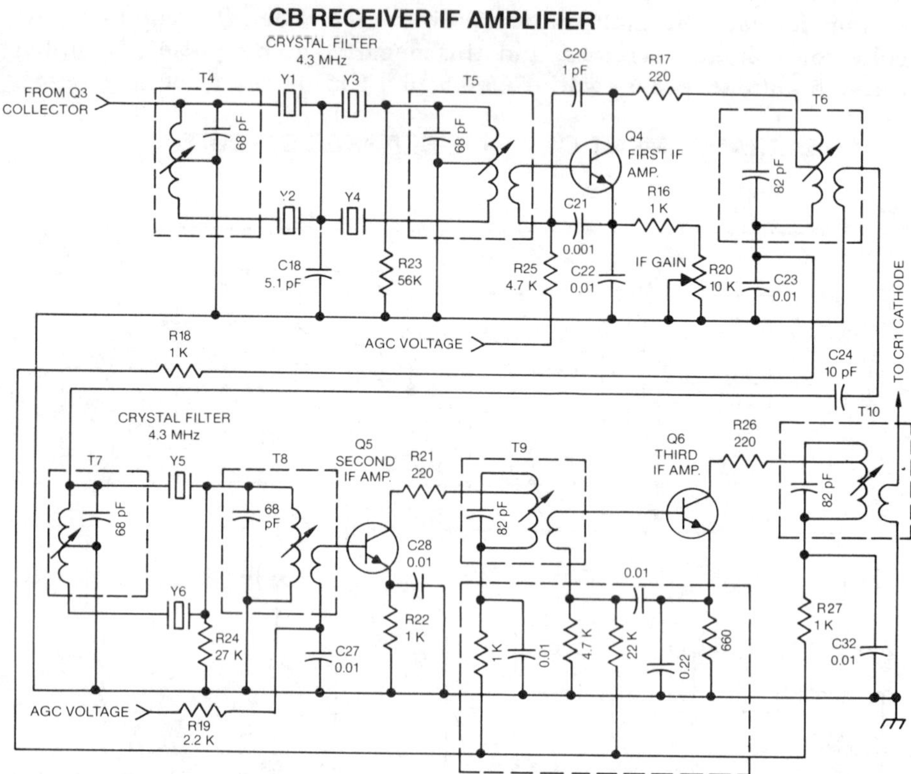

The first IF amplifier amplifies the IF signal to a sufficient level to be filtered again by the second crystal filter. Tuned transformers T7 and T8 match the input and output impedances to provide maximum signal transfer. The crystal filters together with transformers T4 through T10 provide a flat-top, steep-side IF response that establishes the high selectivity necessary to avoid interference from adjacent channels only 10-kHz away. The choice of the 4.3-MHz IF provides good image rejection because of the selectivity of the RF amplifier. While double conversion could have been used to achieve selectivity, the inexpensive crystal filter in the IF provides the same benefits without the additional complication. Provisions are made to apply an AGC voltage to the IF stages to hold the signal level constant over a wide range of input-signal strength.

CB Detector, Noise Limiter, and Audio Switch

The amplified signal from the third IF amplifier Q6 is coupled through IF transformer T10 to CR1, the audio detector shown below. With the IF signal applied, current flows through CR1 only during the negative portion of the IF signal. This causes the 0.047-μF capacitor in the temperature-compensated RC network, U2, to charge to the peak value of the rectified voltage on each negative half cycle. This capacitor and C35 (0.001 μF) shunt the 4.3-MHz component of the IF to ground and effectively filter all IF signals. The time constant of the RC network in CR1 output allows the audio modulation to pass unhindered.

CB RECEIVER DETECTOR, AUDIO NOISE LIMITER, AND AUDIO SWITCH

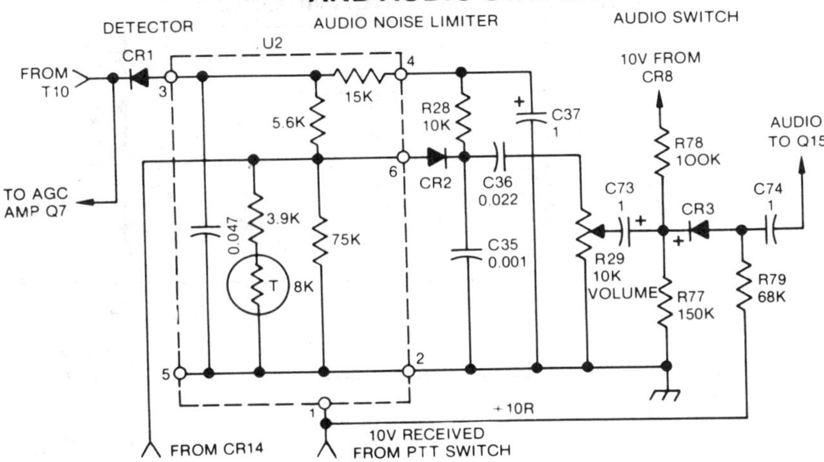

Diode CR2 provides noise limiting. It is forward biased by the 10-volt receive line through the resistive divider in the RC network U2. Bias is set so that CR2 conducts and passes the audio signal developed by CR1 through the 5.6-K resistor in U2, through CR2 and C36, to the manual volume control R29. The RC circuit associated with C37 has a very long time constant, so a dc voltage proportional to the peak carrier level is developed across C37.

A portion of the audio signal discharges C37 and develops a current through R28 to keep CR2 conducting. On noise spikes that exceed the nominal peak carrier (modulation) level, the voltage on C37 does not change, but since the noise spike is negative, CR2 will be reverse biased and stop conducting momentarily on all signals (for example, noise spikes that exceed the peak carrier level). Peak clipping occurs on modulation and noise peaks. This prevents peaks from interfering with audio reception.

After being coupled through C36 to the manual volume control, the signal is coupled through the audio switch diode CR3. This diode is biased on by the 10-volt receive line through R79. Capacitor C74 (1 μF) couples the audio signal to the preamplifier. The audio switch is biased off during transmission by applying the 10-volt transmit-on signal to the cathode of CR3 through R78.

CB Receiver Audio Section

The audio signal from the detector is RC coupled to the base of the audio preamplifier Q15, where it is amplified and fed to the power amplifier U3. C78 and R84 in the collector circuit of Q15 attenuate frequencies above 3 kHz. C81 and R85 form a high-pass filter and attenuate signals below 500 Hz.

EP3 is a ferrite bead, which, when slipped over a wire, increases the inductance of the wire lead. EP3 and C76 act as a filter to prevent any RF signals that may be picked up by the wiring during transmit from being detected by the emitter-base junction of Q15 and appearing as an audio output.

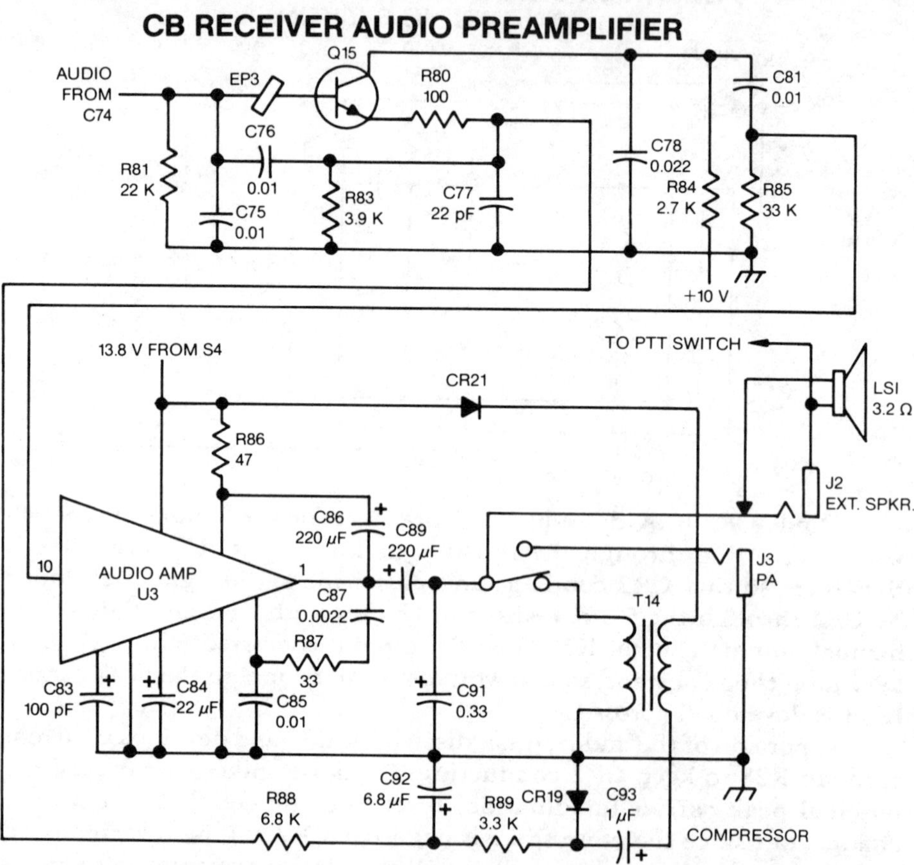

The audio amplifier U3 is an IC linear operational amplifier that produces 4 watts of output power at pin 1 of U3. This output is coupled through the external speaker jack J2 to the loudspeaker. Note that the speaker has an adjacent jack J2 for connecting an external speaker. By using J3, an external power amplifier can be connected to U3. As you learned in Volume 3, this audio power amplifier is also used as the high-level modulator for transmission. Transformer T14 drives the collector circuits of the final power amplifier in the transmit mode.

CB Automatic Gain Control and Squelch

Automatic Gain Control (AGC) is developed by coupling a portion of the signal from the IF output T10 to the AGC detector CR4 and amplifiers Q7, Q21, and Q22. AGC is applied to Q1, Q2, and Q5 in the receiver and Q10 and Q11 in the noise-blanker circuit.

A portion of the amplified AGC voltage is applied to the squelch amplifier Q13 and squelch gate Q14. The squelch gate disables the audio preamplifier by reverse biasing Q15 when there is no *on-frequency* signal at the receiver. The squelch level R73 is adjusted by the operator so as to cut off any signals that are considered to be too weak to be of interest.

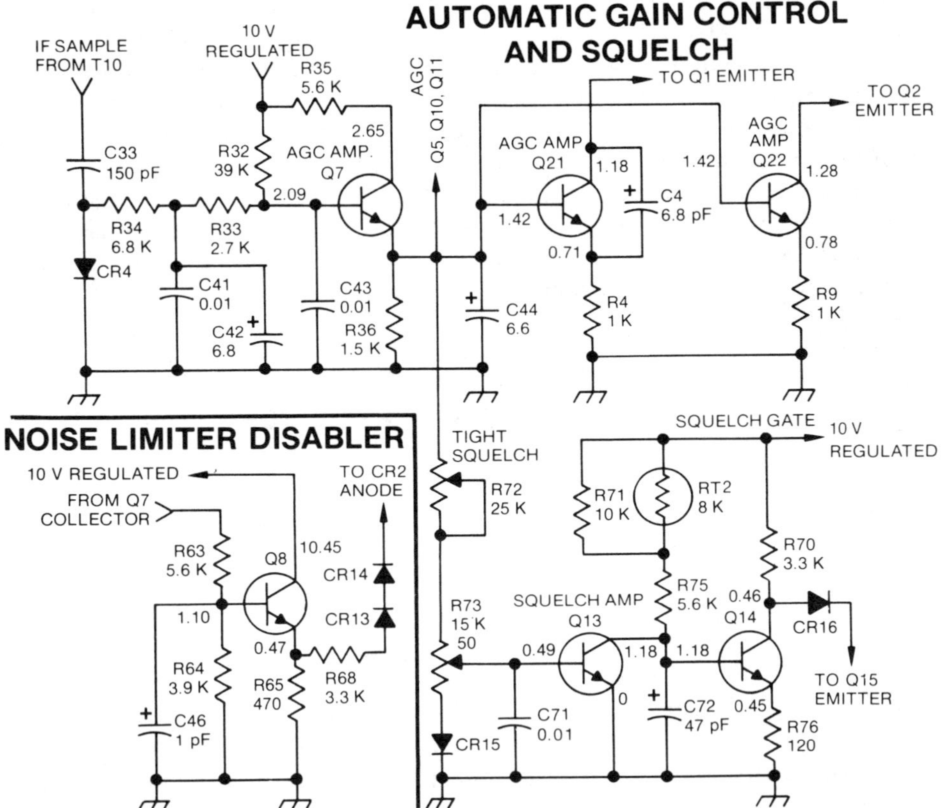

When a sufficiently strong signal is received, AGC voltage is developed on the emitter of Q7, and it is applied to Q13 at the level selected by R73. This causes Q13 collector voltage to increase and forward biases squelch Q14. When Q14 conducts, Q15 conducts and signals pass.

The audio-noise limiter-disabling circuit Q15 places a positive voltage on the anode of the audio-noise limiting diode CR2 discussed previously. This turns on CR2 to prevent clipping and distortion. When a strong signal is received, positive voltage is developed at the collector of AGC amplifier Q7. This turns on Q8, causing CR13 and CR14 to conduct, thus keeping CR2 in conduction—and therefore disabling its limiting function when strong signals are received.

Fleet Communication FM Receivers

Fleet communication receivers serve trucking and taxi fleets and service organizations. They operate in the 150-MHz range and the 450- to 512-MHz region, but other ranges are used as well. In the 450- to 512-MHz range, narrow-band frequency modulation (± 5 kHz) is usually used. These sets generally consist of a base station, which can operate at levels of up to 50 watts, and mobile units operating at power-output levels of up to 15 watts. More than one base station can be used at the same frequency or in the same band, and almost all units are transceivers.

The receiver presented here as an example is similar to the FM receiver described earlier, except for several special circuits. It receives either of two selected crystal-controlled channels at about 450 MHz, although many newer sets use frequency synthesizers. Oscillator frequency-tripling circuits, often used in transmitters, are used to raise the crystal-controlled oscillator frequency to the UHF operating band. A double-conversion superhet receiver is used with a first IF of 10.7 MHz (standard for FM receivers) and a second IF frequency of 455 kHz (standard for AM receivers). This dual-conversion technique is often used in receivers in the VHF–UHF range and above and combines the image rejection of the high-frequency IF with the selectivity of the low-frequency IF.

TYPICAL FLEET COMMUNICATION FM RECEIVER

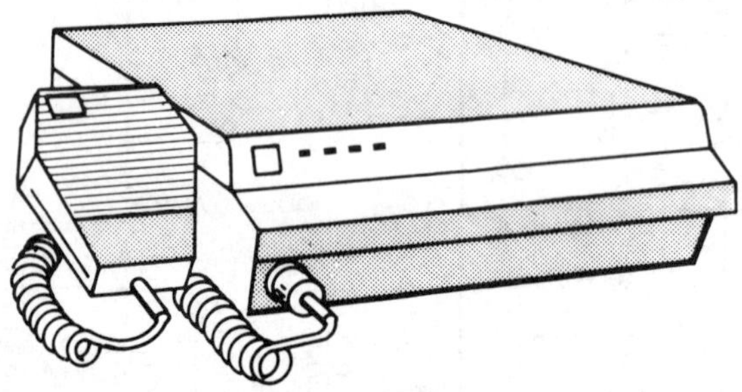

Other special circuits are the squelch filter, squelch gate, and squelch tail eliminator. The squelch circuits keep the receiver audio amplifier turned off until a carrier is received. This avoids the necessity of listening to receiver noise when no signal is present. Another special feature of this system is the *call guard*. When no communication is being received, the receiver audio output is turned off. A special call-guard tone is transmitted with the message intended for a specific receiver in a network or net of receivers. When this tone is received, the receiver audio output is turned *on*, and the message can be heard. With the squelch and call guard operating, only messages intended for a particular unit or group of units will be received.

Fleet Communication FM Receivers (continued)

The figure below is a block diagram of the receiver portion of the E. F. Johnson Fleetcom® FM transceiver. You will recall that you studied the transmitting portion of this system in Volume 3. As shown, the signal input from the antenna is filtered in a bandpass filter to reject undesired signals and the image signal. An RF amplifier is omitted in this receiver to reduce cost. The output of the bandpass filter feeds the first mixer directly. The first local-oscillator signal is crystal controlled and produces a first IF signal at 10.7 MHz to the high-frequency IF amplifier. This provides good image rejection. The amplified 10.7-MHz signal is then converted to 455 kHz via the second mixer and second local oscillator. This 455-kHz signal is then amplified to provide high selectivity, (better than that readily obtainable at 10.7 MHz). The output signal from the second IF is applied to a limiter/discriminator to produce an audio output.

SIMPLIFIED BLOCK DIAGRAM OF A FLEETCOM II FM RECEIVER

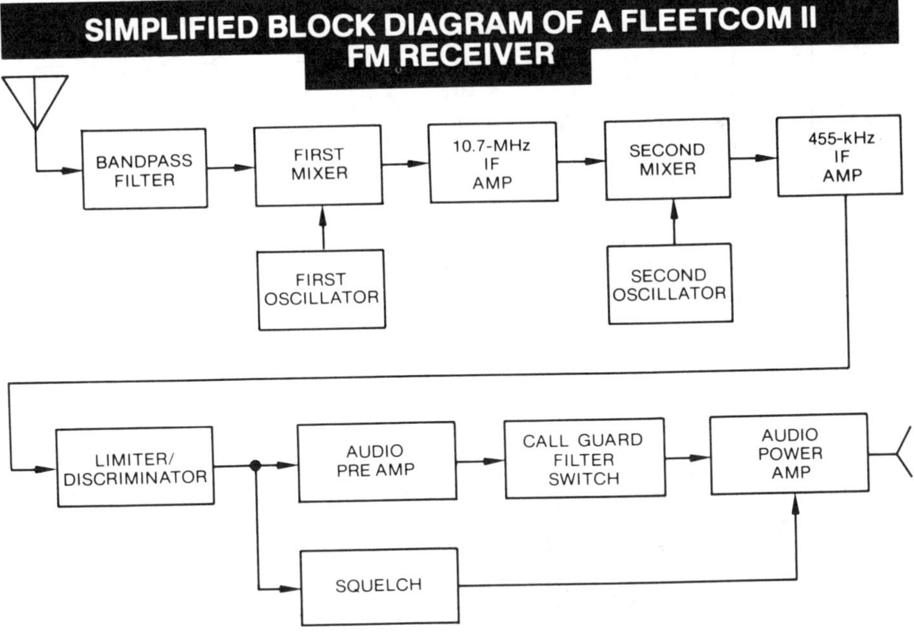

The audio signal from the discriminator is preamplified and fed to the audio amplifier. The preamplifier output is also applied to the call-guard filter. The audio output is restricted if the call-guard tone is absent, even though a signal may be present. A call-guard filter in the audio line suppresses the tone during reception. The squelch line operates from the discriminator output. When the signal level exceeds the manually set squelch threshold, the audio amplifier is enabled to allow the signal through. Both the squelch and call guard can be disabled by the operator when desired.

Fleet Communication RF Filter/First Mixer/First Oscillator/ Triplers

As shown below, the antenna relay couples the antenna to the receiver input when in the receive mode. The incoming RF signal at about 450 MHz is coupled through antenna relay K1 and a bandpass filter, tuned by C101, C102, C103, C104, and C105, to the first mixer. The bandpass filter is wide enough in response to receive either channel without retuning but narrow enough to avoid the image signal 21.4 MHz away.

The first mixer, Q103, mixes the RF input signal with the output of the third local-oscillator tripler (Q101). The first mixer uses a MOSFET described earlier. The first oscillator frequency is chosen to produce an IF signal at 10.7 MHz. The mixer output is then coupled to the 10.7-MHz IF amplifier, Q201, through a conventional IF input transformer tuned to 10.7 MHz.

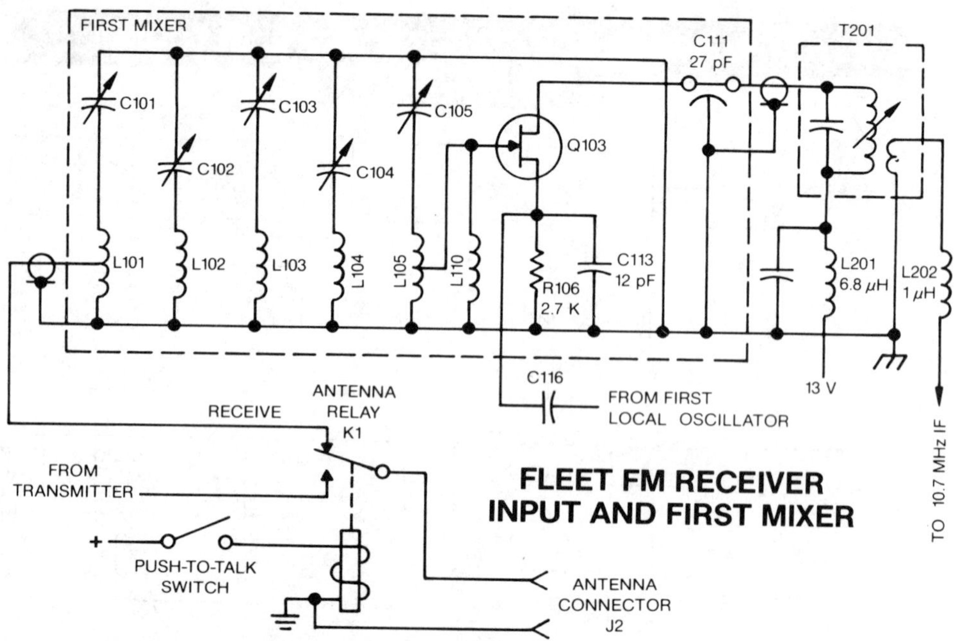

FLEET FM RECEIVER INPUT AND FIRST MIXER

The figure on the following page is a diagram of the first oscillator. Channel selector switch S2 grounds the emitter resistor of the appropriate channel oscillator—Q211 for channel 1 or Q212 for channel 2. Both oscillators are crystal controlled using a modified Colpitts circuit. Their outputs are tuned to the third harmonic of the crystal, and they function as frequency triplers. Transformers T208 and T209 provide the output tuning. A passive temperature-compensating technique is used with RT201 to control the effective capacitance of C291. The frequency is adjusted by C269 and can be extended by changing the value of capacitor C271.

FLEET FM RECEIVER

FIRST OSCILLATOR AND FREQUENCY TRIPLERS

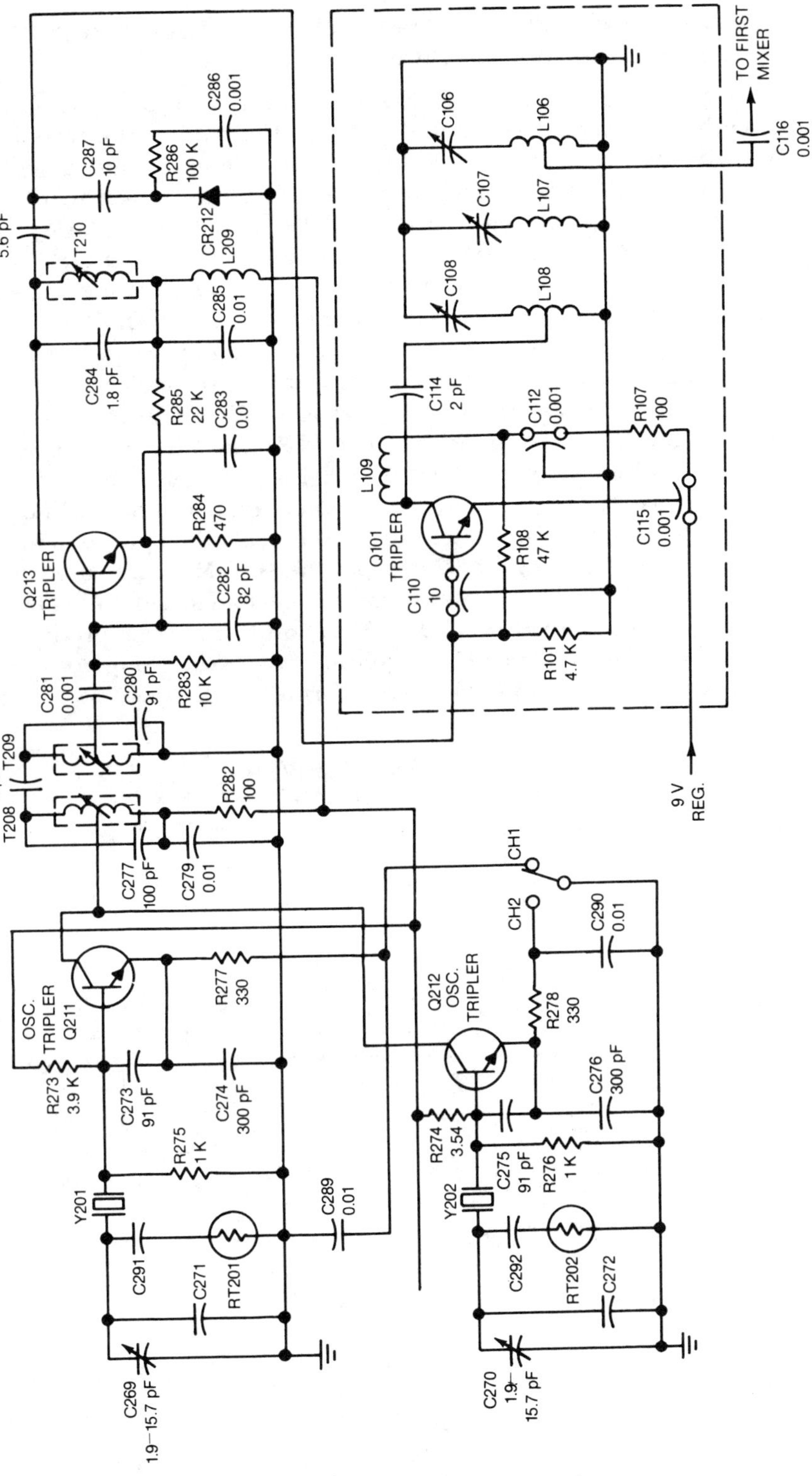

C286 5.6 pF
C287 10 pF
R286 100 K
C286 0.001
CR212
T210
L209
C284 1.8 pF
C285 0.01
R285 22 K
C283 0.01
Q213 TRIPLER
R284 470
C282 82 pF
C281 0.001
C280 91 pF
R283 10 K
C278 3 pF
T209
T208
R282 100
C277 100 pF
C279 0.01
OSC. TRIPLER Q211
R273 3.9 K
C273 91 pF
C274 300 pF
R277 330
R275 1 K
Y201
C291
RT201
C271
C289 0.01
C269 1.9–15.7 pF
CH1
CH2
C290 0.01
Q212 OSC. TRIPLER
R278 330
C276 300 pF
R274 3.54
C275 91 pF
R276 1 K
Y202
C292
RT202
C272
C270 1.9–15.7 pF
9 V REG.
TO FIRST MIXER
C116 0.001
C106
L106
C107
L107
C108
L108
C114 2 pF
C112 0.001
R107 100
L109
Q101 TRIPLER
C115 0.001
R108 47 K
C110 10
R101 4.7 K

Fleet Communication First Oscillator and Triplers (continued)/ First IF Amplifier and Second Mixer

On the previous page, the tripled output of the oscillator is coupled through transformers T208 and T209 to the base of the second tripler Q213 through capacitor C281. The second tripler output, tuned to nine times the oscillator frequency, is coupled through transformer T210 and capacitor C286 to the base of the third tripler Q101. The signal from the third tripler (27 times the crystal frequency) is filtered by C106, L106, C107, L107, C108, and L108 to remove the undesired harmonics. The filtered signal is coupled to the source of the MOSFET first mixer Q103 through C116. The crystals have fundamental frequencies (in megahertz) equal to the channel frequency ±10.7 MHz/27.

As shown below, transistor Q201 is a FET amplifier tuned to 10.7 MHz. It amplifies the signal at 10.7 MHz prior to conversion to the second IF. The output is tuned by L204 and coupled to a crystal filter Z201 by capacitor C204. Z201 has a center frequency of 10.7 MHz and a bandwidth of 13 kHz. Matching to the second mixer is adjusted by L205. The second mixer Q202 accepts the 10.7-MHz signal output from the crystal filter and the output of the second oscillator. The second local oscillator is tuned to 11.155 MHz, and the mixer produces a difference frequency of 455 kHz. The mixer output is coupled to the low-frequency IF amplifier Q203 by transformer T202 tuned to 455 kHz.

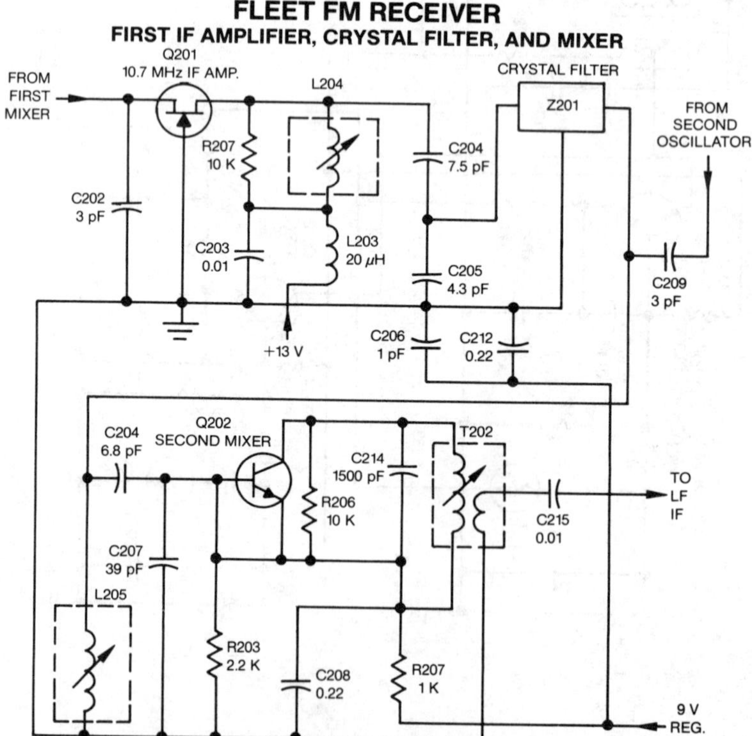

Fleet Communication Second Oscillator and IF Amplifiers

As shown in the schematic diagram below, the second oscillator Q214 operates as a parallel-mode Colpitts oscillator with the frequency controlled by crystal Y203 and positive feedback controlled by C263 and C264. The crystal operates as a parallel-resonant element and oscillates at 11.155 MHz (or 10.245 MHz as an optional *low-side* injection frequency). The signal from the second oscillator Q214 is coupled to the base of the second mixer Q202 through capacitor C209.

The second mixer output, containing the original frequencies plus the sum and difference of these frequencies, is tuned to the difference frequency, 455 kHz, by the tuned IF transformer T202. The signal is amplified by the two common-emitter IF amplifiers Q203 and Q204. The output of the second IF amplifier is coupled to the limiter/discriminator IC U201 by IF transformer T204.

FLEET FM RECEIVER SECOND OSCILLATOR AND 455kHz IF AMPLIFIER

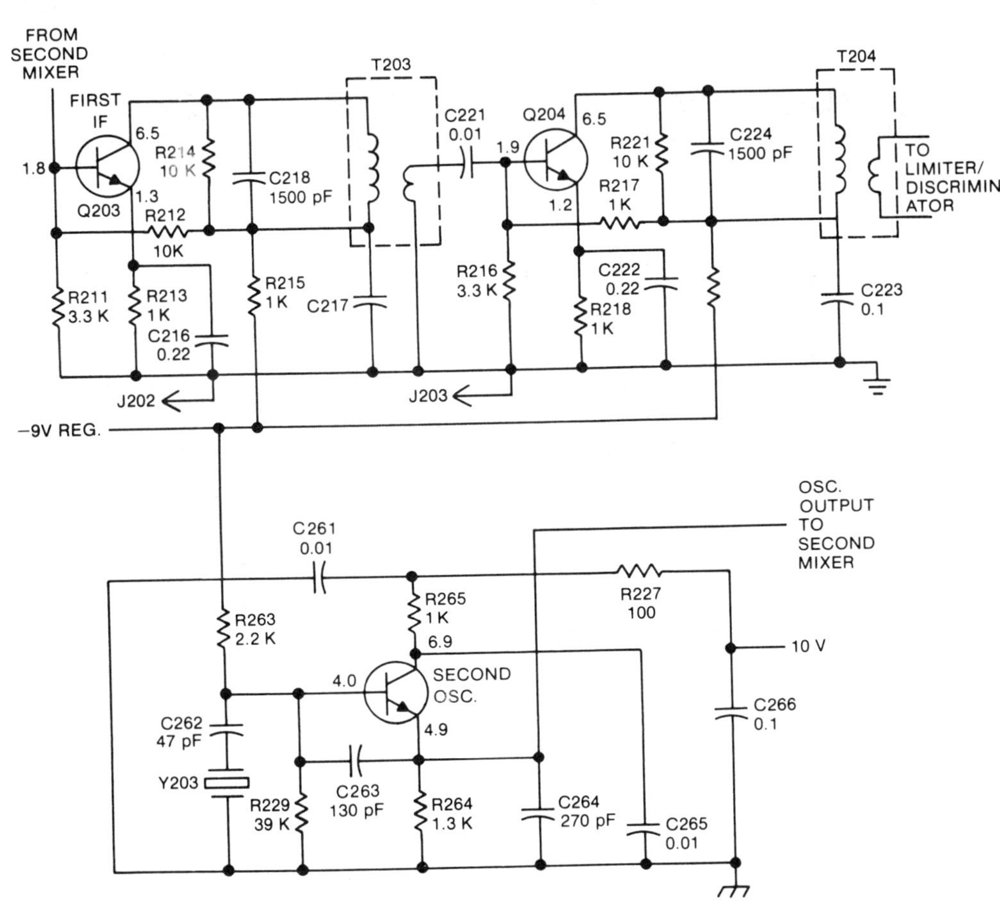

Fleet Communication Limiter/Discriminator and Audio Preamp

As shown in the schematic diagram below, U201 is a monolithic IC that includes a three-stage FM IF amplifier/limiter, a doubly balanced quadrature FM detector, and an audio preamplifier. The limited IF signal is coupled from pin 8, through C232, to the 455-kHz tuned network (C239/T205). This tuned circuit is external to the IC because of its size. A signal is fed into pin 9, 90 degrees out of phase with the original IF signal, which is combined with the normal (in-phase) IF signal in a double-balanced mixer in the U201 FM discriminator. The audio modulation (recovered as you learned in your study of FM detectors) is amplified in U201 and then fed to the de-emphasis network C230, C237, C255, R232, R233, and R234—which provides about 12 dB per octave noise reduction above 3 kHz.

The de-emphasis network plus feedback network C247 and R231 across amplifier U202A provide approximately 6 dB per octave de-emphasis from 700 to 3,000 Hz and compensate for pre-emphasis provided in the transmitter. The amplifier U202A provides an audio output for the audio power amplifier. A squelch circuit, similar to that described earlier for the CB transceiver, provides for audio silence when no signal is received.

FLEET FM RECEIVER LIMITER-DISCRIMINATOR AND AUDIO PREAMPLIFIER

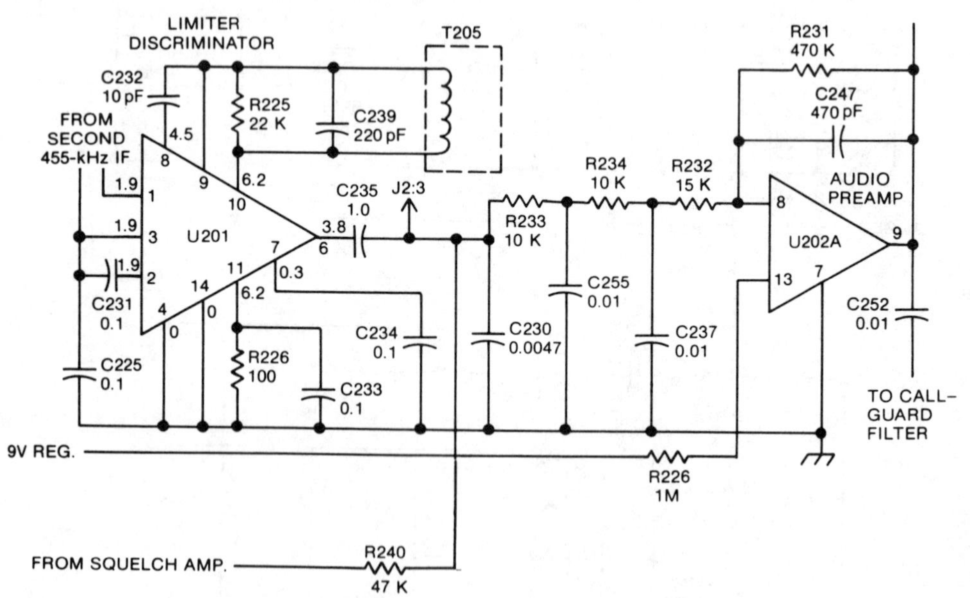

Fleet Communication Audio Section

The *call-guard* feature of the transceiver uses low-frequency audio tones to activate this feature. These are removed from the audio signal by the call-guard filter. The *call-guard* filter U202B functions as a high-pass filter with a low-frequency cutoff of approximately 340 hertz to remove the *call-guard* tone from the receiver audio output. The output of this stage is coupled to the manual volume control.

The audio signal passed by the call-guard filter is fed through the volume control R241, amplified by audio amplifier Q206 and audio driver Q207, and then coupled by transformer T206 to the audio output stage.

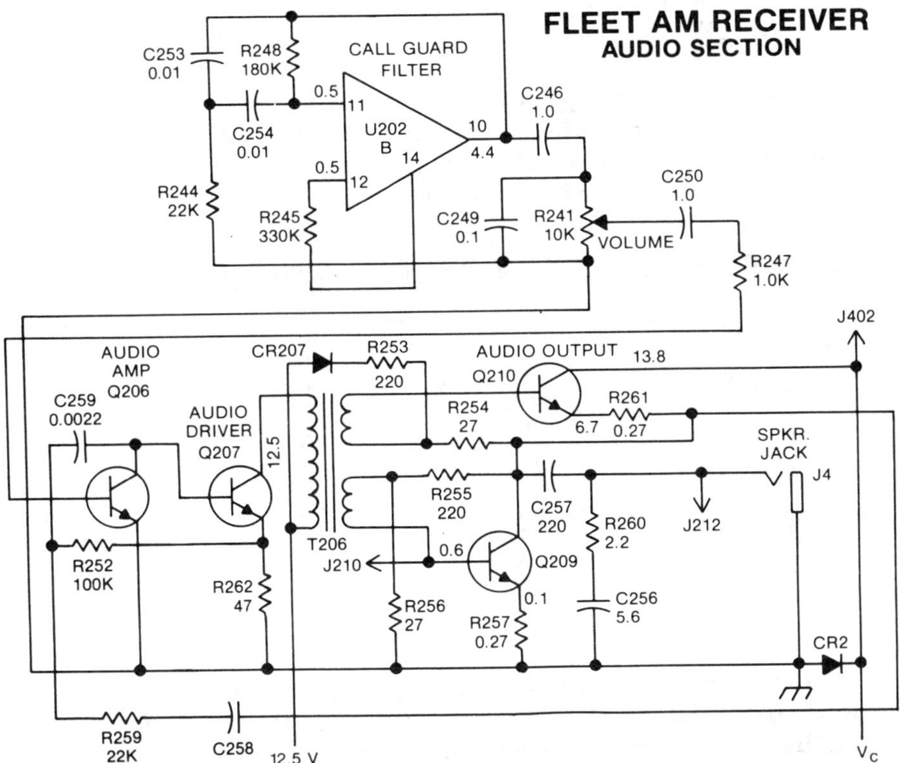

Transformer T206 provides the split phase required to drive push-pull output amplifiers Q209 and Q210. Operating bias is developed by R254 and R256 for Q210 and Q209, respectively. A dc bias stability is provided by R257 and R261.

Negative-feedback network C258 and R259 provides for better linearity and frequency response of the audio amplifier, which results in better sound quality. Capacitor C258 couples the amplified audio signal to the speaker jack. Series network C256 and R260 prevents oscillation caused by the inductive effects of the speaker voice coil.

Amateur Radio/Short-Wave Receivers

Amateur radio is very popular in many countries and offers a way for individuals to communicate by radio and to experiment with transmission and reception. Radio amateurs have made many important contributions to radio and communications technology. The amateur-radio frequency bands are distributed over the spectrum, with the more important ones lying between 1,800 kHz and the UHF region. The table below shows the lower frequency bands that are most popular:

1.8 - 2.0 MHz	28.0 - 29.7 MHz
3.5 - 4.0 MHz	50 - 54 MHz*
7.0 - 7.3 MHz	144 - 148 MHz*
14.0 - 14.35 MHz	220 - 225 MHz*
21.0 - 21.45 MHz	*USUALLY INDIVIDUAL RECEIVERS.

To accommodate this range of frequency coverage in a single receiver requires a method for rapidly changing from one band to another. The band-switching receivers use a selector switch in the RF section to change the inductors to allow for band switching, as shown.

BANDSWITCHING FOR WIDE FREQUENCY COVERAGE

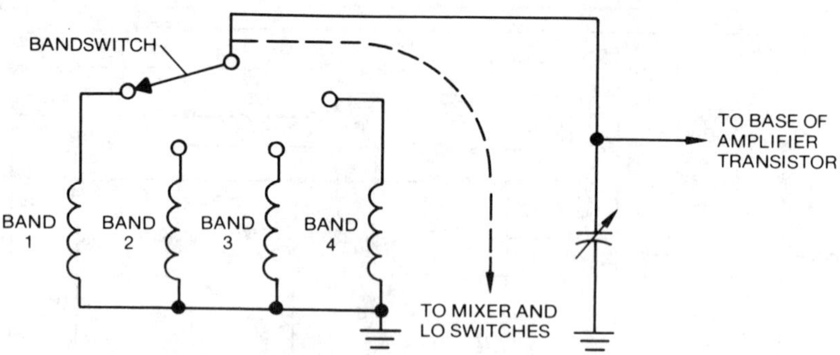

Similar inductor/switch arrangements are used at the mixer input and in local-oscillator circuit to switch these appropriately at the same time. Usually, these switches are ganged together so that a single control is used to switch rapidly and conveniently from band to band. In some cases, the band coverage is continuous; that is, the receiver bands are arranged so that the next band begins where the previous band left off. In this way, the entire short-wave range from the broadcast band through the Citizens Band and higher can be covered in a single receiver. These receivers are used to listen to short-wave broadcasts and communications. Usually, special filter circuits in the IF allow for adjustment of the IF bandwidth to match the signal bandwidth so that interference can be minimized. A BFO (beat-frequency oscillator) tuned a few hundred hertz from the IF can be introduced at the second detector. This allows CW signals to be converted to tones by the mixing action at this second detector, which produces a beat-frequency output between the IF and the BFO.

Receivers for Public Service/Navigation/Communications/Weather

There are many specialized navigation systems that use the low-frequency ranges below the standard broadcast band because of their stable propagation characteristics. In addition, there are ship emergency frequencies (for example, 500 kHz) used internationally for distress calling. The VHF–UHF range is used for any number of specialized public services. For example, the VHF region slightly above the FM broadcast band is used for air-to-air and air-to-ground communication and navigation. The VHF frequency range between 148 and 174 MHz is used extensively for police and fire department communication, as well as the UHF bands. These bands are also used by boats for intercommunication and for communication with the U.S. Coast Guard in coastal waters.

Of particular interest is the weather broadcasts in the 170-MHz range provided by the U.S. Weather Service. These broadcasts are continuous and provide instant access to local weather as well as weather forecasting. The weather information is provided on a 24-hour basis so that all within range of these stations can get weather information as needed. While these are very useful to individuals who own a suitable receiver, they are very important to ships and pleasure boats in coastal waters since they provide information on storms and other marine conditions important to the boat operators.

The National Bureau of Standards radio stations WWV provide transmissions at 5, 10, 15, and 20 MHz. These stations not only provide very accurate radio transmission for equipment calibration, but also provide time signals and special information via propagation.

All of these systems use receivers that do not differ materially from the receivers you have already studied. The only differences lie in the *antenna system* and the *tuning range* of the RF portion of the receivers.

Satellite Systems

As you know, radio-frequency propagation at VHF and above is pretty much limited to line of sight. On the other hand, the available spectrum space to carry the increasing load of information is severely limited at longer wavelengths. Furthermore, except at low frequencies, operation in the high-frequency range is often unreliable. Satellite relays solve the line-of-sight problem very neatly and allow the use of the spectrum space available at VHF and above. Nowadays, most communication satellites are in synchronous orbits—that is, they stay in one spot relative to a point on the earth. The altitude for a synchronous orbit is about 22,000 miles. While it may seem that this is doing things the hard way, it is important to realize that *only three satellites* can provide for communication to *any point on the earth*, including ships at sea and aircraft!

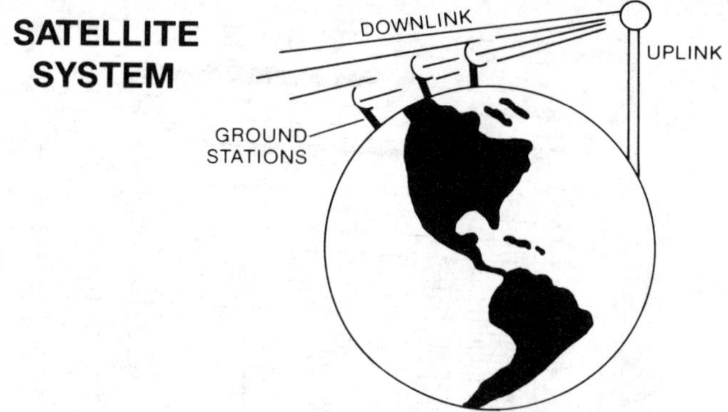

Modern communication satellites are really complex relay stations. They receive transmitted signals from appropriate ground transmitting stations, convert these to signals at a different frequency, amplify the signals, and rebroadcast these to appropriate ground receiving stations. Both AM and FM transmission are used. Since highly directive, high-gain antennas can be used both for transmitting and receiving at the ground station end, the transmitter power required at the ground transmitter and in the satellite is not great.

A wide variety of information is sent via satellites, including not only communications such as telephone calls but also TV programs, radio programs, the facsimile editions of entire newspapers, digital data, or any other information that is desired. The use of satellite relays is increasing, and nowadays they are replacing other means of relay for long distances.

Satellites are owned either by governments or by corporations. Because of the great expense of building and orbiting a satellite, many corporations owning satellite facilities are actually formed by several companies to share the cost and the facilities. These special-communication corporations then rent use of these facilities just as land communication facilities (wire and coax cable) are rented for use by private corporations.

Single-Sideband Suppressed-Carrier Receivers

Single-sideband (SSB) transmission is widely used for communication, as you learned in Volume 3 on transmission systems. The basic differences between a single-sideband suppressed-carrier receiver and a conventional receiver are (1) increased selectivity because only one sideband is to be received and (2) the carrier must be reinserted at the detector for the detection process to take place. The E. F. Johnson Company Viking 4740 is a CB transceiver that has a single-sideband transmit and receive mode. Refer to Volume 3 for the details of the SSB transmitter portion. As you may remember, in this transceiver a crystal filter is used to suppress the unwanted sideband, while the carrier is suppressed by a balanced modulator. The same filter is used on receive to filter out unwanted interfering signals on the unused sideband. Thus, the filter FL501 is multiplexed for both transmit and receive, but is used in the same way for both. You will recall that the transceiver uses a frequency synthesizer to produce a local-oscillator output signal 7.8 MHz below the transmitted frequency.

FILTER AND DETECTOR SYSTEM FOR SSB RECEPTION

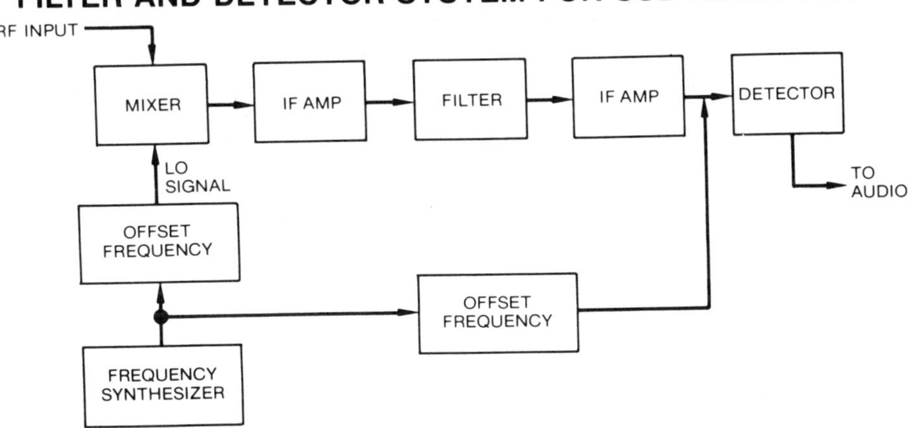

The figure shown is a partial block diagram of the filter and the detector system for SSB reception. Rather than use two filters [one for the upper sideband (USB) and one for the lower sideband (LSB)], a single filter is used, and the local-oscillator frequency is shifted slightly to allow for reception of either sideband. When this is done, the injected detector carrier must also be shifted; otherwise the sideband data for one case will also be inverted. With these exceptions, the transceiver circuits are essentially identical to the transceiver receiving circuits you studied earlier. The table shown below indicates the frequencies used for DSB, USB, and LSB transmission and reception.

MODE	CARRIER FREQ	SYNTHESIZER OUTPUT FREQ (LO)	IF SIGNAL BAND	DETECTOR CARRIER
DSB	f_o	$(f_o) - 7.8$	7.7975 – 7.8025	
LSB	f_o	$(f_o - 0.0025) - 7.8$	7.7975 – 7.8025	7.8025
USB	f_o	$(f_o + 0.0025) - 7.8$	7.7975 – 7.8025	7.7975

Single-Sideband Suppressed-Carrier Receivers (continued)

You can see how DSB, USB, and LSB reception can take place using a single filter. For example, if a filter centered at 7.8 MHz covering the band from 7.7975 to 7.8025 MHz (±2.5 kHz) is used, it can provide for selectivity for DSB reception by centering the carrier at 7.8 MHz, as shown, so that both sidebands are passed by the filter. To provide lower sideband only, we offset the carrier by 2.5 kHz to 7.8025 MHz. The lower sideband is now located in the filter passband. For the case shown, the carrier is reinserted at 7.8025 MHz to retain the correct relationship between the sideband components. Similarly, the upper sideband can be selected by shifting the local oscillator (or transmit signal) down to 7.7975 MHz so that the sideband lies in the filter passband.

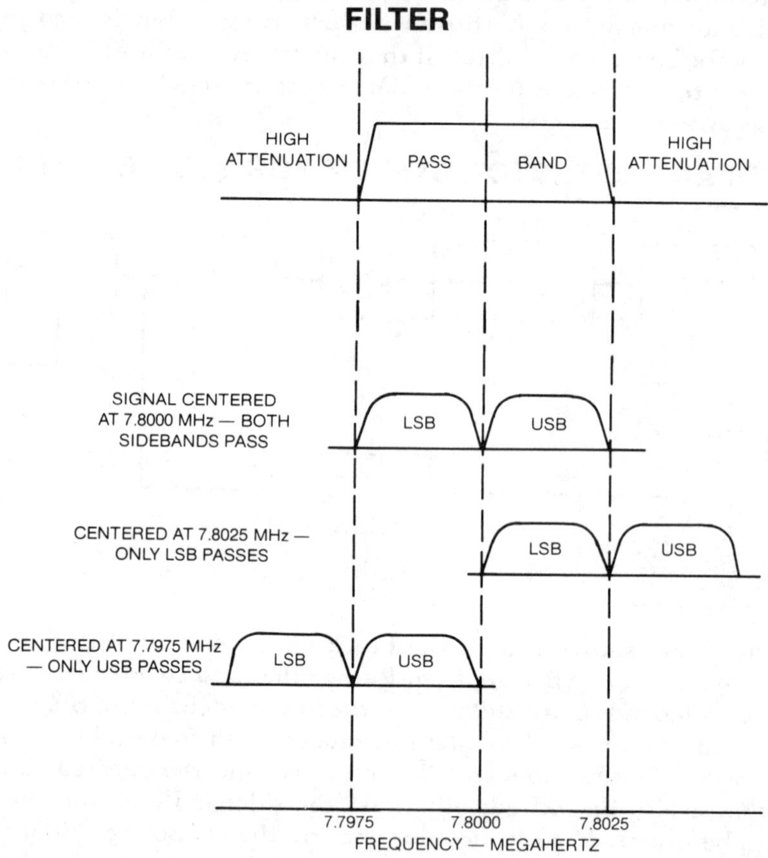

As you can see from the figure, the choice of upper or lower sideband reception is accomplished with a single filter by simply shifting (offsetting) the IF carrier as shown. Note that the proper frequency for detector injection must be used so that the sideband does not become inverted.

Digital Communications Receivers

The principles of receiver operation that you learned about previously are directly applicable to receivers designed to receive digital signals. Thus, a conventional AM or FM superhet receiver that you learned about earlier can be used for reception of digital or pulsed signals. The only difference lies in the *bandwidth* of the receiver circuits. You will learn about the nature of the digital signal and its information content when you study digital circuits in Volume 5. You will also remember from Volume 3, the study of transmission systems, that the bandwidth required for a pulsed signal is about equal to the reciprocal of the pulse width. Thus, 1-μsec pulses require an overall receiver bandwidth of about 1-MHz in the RF/IF portion. A bandwidth of half this is sufficient for the detector/video amplifier portion. Since bandwidth is related to tuned circuit Q, the Q must be reduced if it is too high.

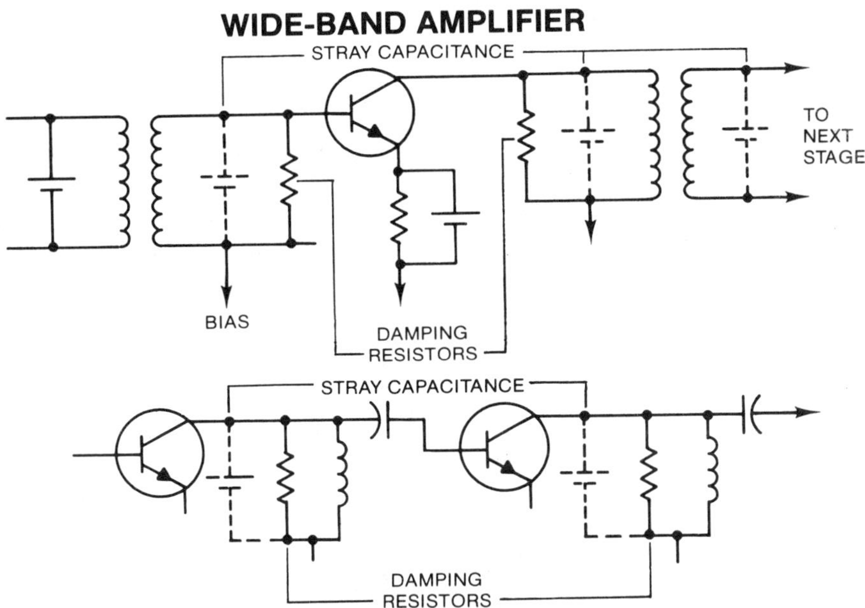

The usual configuration for wide-band amplifiers uses capacitance coupling and single tuned circuits, as shown. Since $Q = \omega L/R$, the damping resistor serves to reduce the Q of the circuit. As you may remember, Q is also defined as $\Delta f/f_o$, where Δf is the bandwidth. Therefore reduced Q means greater bandwidth. In the following study of TV receivers, you will see that damping resistors are used on the tuned circuits to reduce the Q by the desired amount. In fact, TV receivers are good examples of pulse or digital signal receivers, since they must deal with signals that are pulse-like in their characteristics.

As you can see, the only difference between a digital or pulse receiver and a conventional receiver lies in the bandwidth of the RF/IF system and bandwidth of the amplifiers (video) following the detector.

Review of Communication Receivers

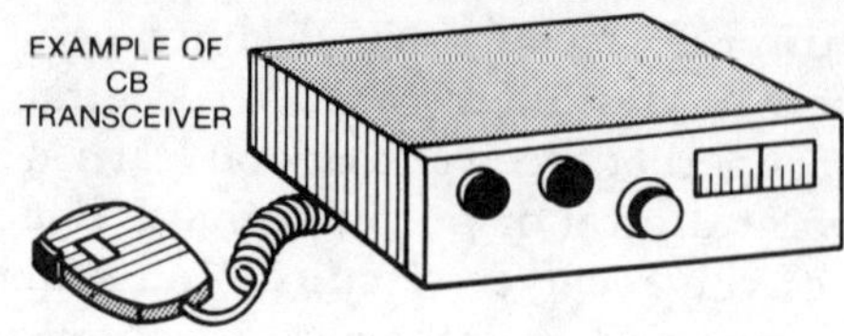

1. COMMUNICATION RECEIVERS are for receiving information as opposed to entertainment. They are most often parts of *transceivers*.

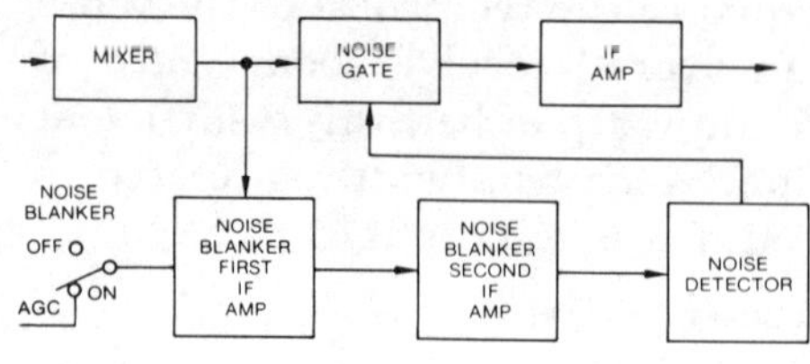

2. CB RECEIVERS feature special techniques such as synthesizer frequency selection and *noise limiting*.

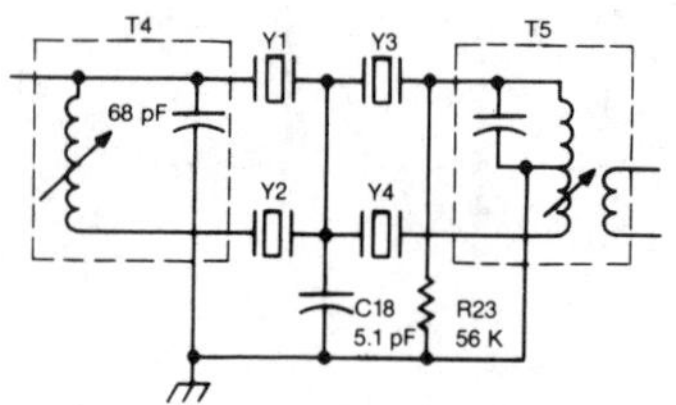

3. IF RESPONSE AND SELECTIVITY are other features of CB receivers. To minimize interchannel interference, *ceramic crystal filters* are most often used.

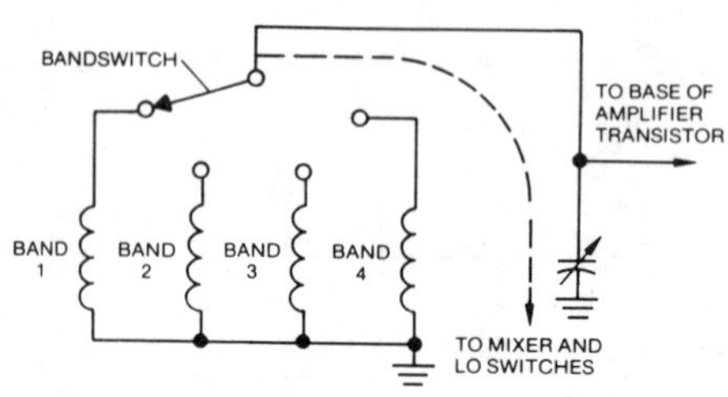

4. FLEET COMMUNICATION AND AMATEUR RECEIVERS are other communication types. Amateur receivers must cover a wide range of frequencies and do it by use of *band switching*.

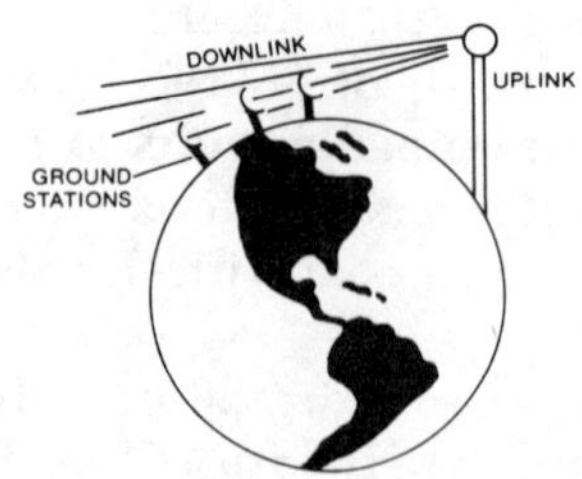

5. SHIP EMERGENCY, WEATHER, AIR-TO-GROUND, SATELLITE COMMUNICATION, and accurate frequency and timing broadcasts are some of the other transmissions for which receivers are used.

Self-Test—Review Questions

1. What is a transceiver?
2. What modes of transmission/reception are permitted for communication systems?
3. Draw a block diagram of a CB-receiver/digital-frequency synthesizer. Describe how it operates.
4. Why are synthesizers desirable?
5. Compare single- and double-sideband reception.
6. What are some of the essential differences between a broadcast receiver and a communication receiver?
7. What do most short-wave receivers use to cover a wide range of frequencies?
8. What are some of the special broadcast services available from WWV operated by the U.S. federal government?
9. Why would anyone want to use a satellite communication system?
10. What is the difference between a pulse or digital receiving system and a conventional voice system?

Learning Objectives—Next Section

Overview—You will next proceed to a study of the television system as a whole. This will review what you studied about TV in Volume 3 and prepare you for TV receiver details that follow.

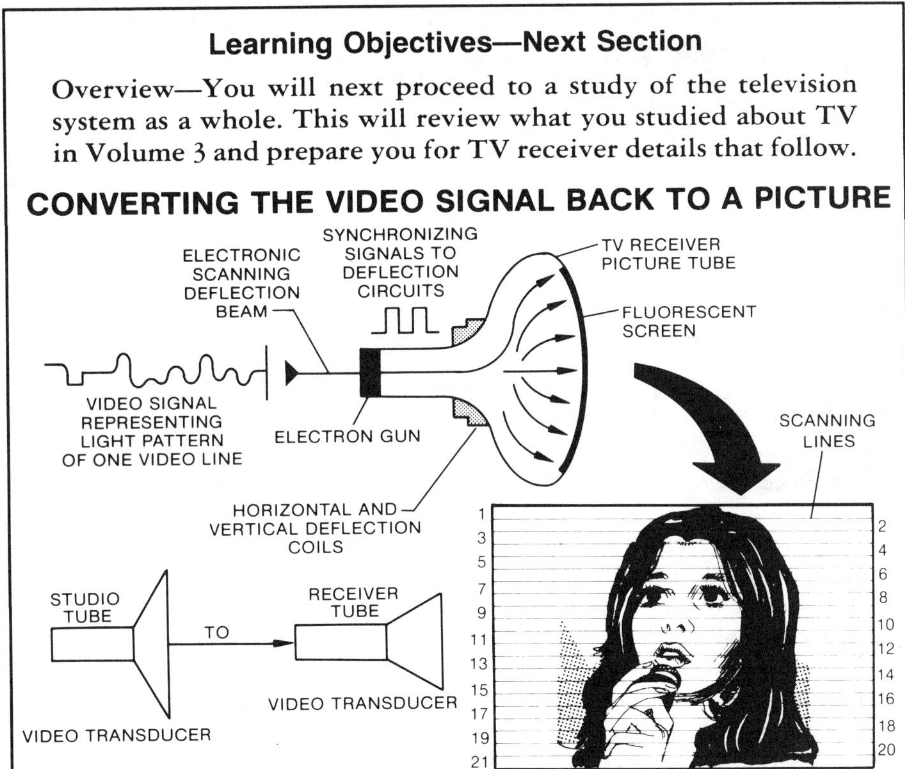

The Television System

Television (*distant vision*) is an electronic system that conveys both *visual* (video) and *aural* (audio) *information* from one location to another. When these locations are *far apart*, the process includes the impression of the video and audio information on an RF carrier and radiation through space. When the transmission and reception locations are *nearby*, the information can be carried, at baseband, over wires from the camera/microphone to the receiver/monitor. This is called *closed circuit TV (CCTV)*. *Cable TV (CATV)* is also very popular. In CATV, a master receiver system is used to pick up and amplify transmitted signals, and these are distributed to conventional TV receivers via coaxial cable.

THE TV TRANSMISSION AND RECEPTION PROCESS

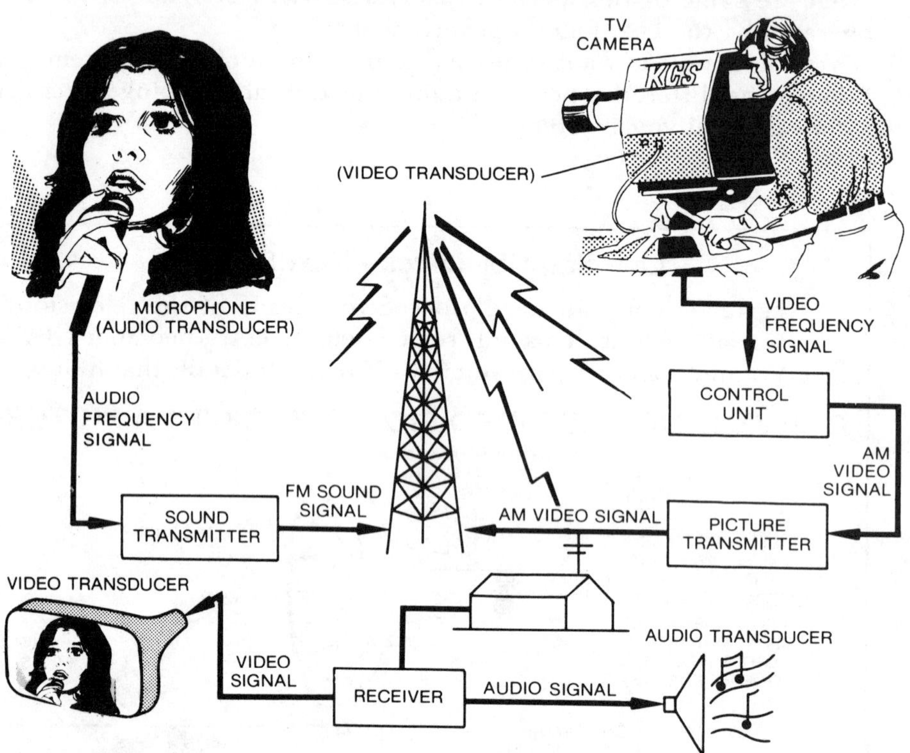

In the original scene, the sound information is converted to audio and used to frequency modulate the aural RF carrier. The TV camera scans the scene and converts the visual information to a video electrical signal. The video is combined with blanking and synchronizing signals and used to amplitude modulate an RF carrier that is 4.5 MHz lower in frequency than the sound carrier. The transmitted AM and FM signals are received using AM and FM receiving techniques with which you are now familiar. What is *new* in TV reception is *how the AM video signal is processed to recover the synchronizing/blanking signals and the visual information and uses them to provide a duplication of the original scene on the receiver picture tube.*

Types of TV Systems—CCTV and CATV

As you learned in Volume 3, the video and audio (aural) information signals are impressed on RF carriers for transmission through space. If the signals are used directly *without* going to RF, the system is called *closed circuit TV* (CCTV) and the connection is made via coaxial cable. In this case no RF is involved, and the video and audio signals are sent *directly* on the cable. When this is done, *no transmitter* is involved and the RF, IF, and detection portions of the receiver are not needed. The *TV monitors* used in CCTV are essentially TV receivers containing *only the video and CRT circuits* so that the input signals at baseband (video and audio) are used *directly*, as will be described later. TV systems of this type are widely used for security systems and monitoring of industrial processes. In some cases, the video and audio signals are converted to low-power RF signals (in a converter that is actually a miniature transmitter) so that a conventional TV receiver can be used at the receiving end. This is done because conventional TV receivers are readily available and, strangely enough, less expensive than TV monitors because conventional TV receivers are produced in such great quantities.

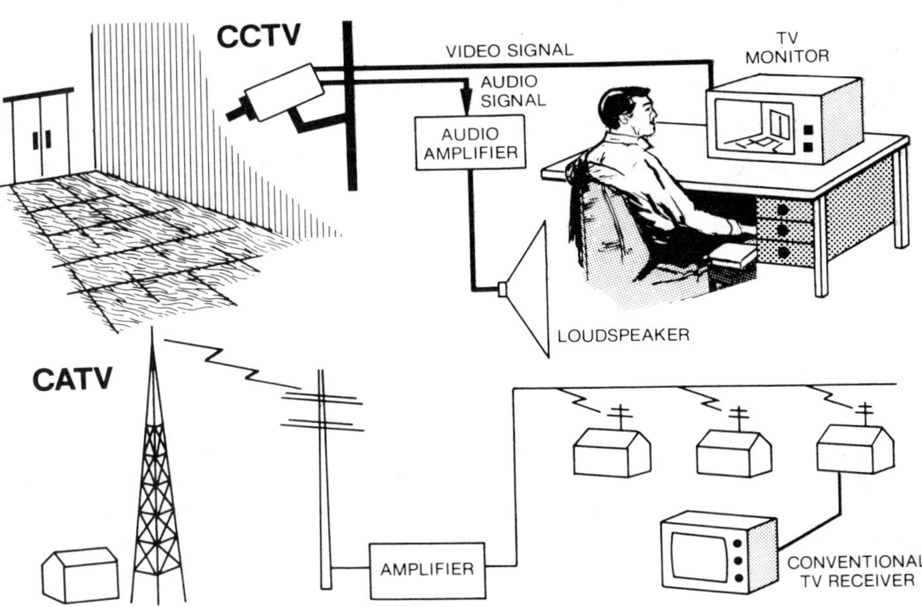

CATV or cable TV was originally designed to provide TV to homes in areas where reception was poor. A large, complicated, and expensive antenna system was erected, and the received signals were amplified and distributed at RF to homes via a coaxial cable. Nowadays, these systems are in use almost everywhere, even in good signal areas, because they have been expanded not only to provide excellent reception of local stations but also distant stations and *special programming* of movies, sports events, and local community events. These special programs are available only to CATV subscribers.

TV Frequency Allocations

The audio (sound) portion of a TV program is transmitted monaurally by frequency modulating a separate RF carrier specifically for sound transmission. For this purpose, an RF bandwidth of 50 kHz (±25-kHz deviation) is adequate. In contrast, the video signal required for good picture definition has frequency components of up to 4 MHz and higher. As you know, to broadcast this signal by conventional double-sideband methods would require a double-sideband bandwidth of 8 MHz plus additional unused bandwidth to separate it from its own sound channel and from adjacent TV channels. To allocate such a bandwidth to a large number of TV channels would use up too much of the radio spectrum. The method finally established in the United States as standard is to use a combination of double- and single-sideband transmission known as *vestigial-* (or part-) *sideband transmission.* A 4.0-MHz bandwidth is assigned to the upper sideband, but only a 1.25-MHz bandwidth is given to the lower sideband. This provides good picture quality and saves bandwidth. It requires, however, a special design of the receiver IF stages.

U.S. VESTIGIAL SIDEBAND TRANSMISSION

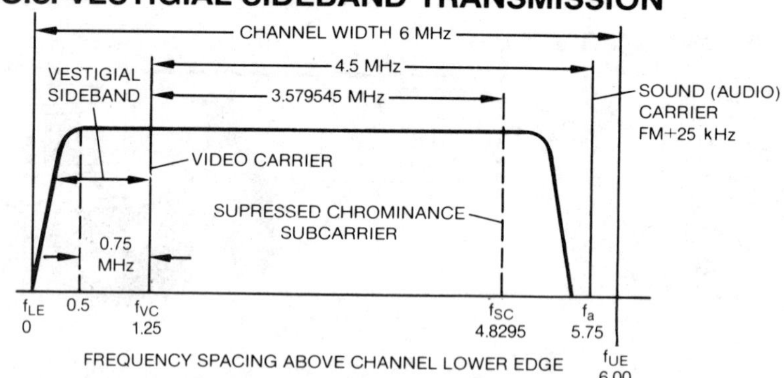

f_{LE} = FREQUENCY LOWER EDGE OF CHANNEL (REFERENCE)
f_{UE} = FREQUENCY UPPER EDGE OF CHANNEL

Each TV channel is assigned a 6-MHz bandwidth in which the sound RF carrier is at a frequency 4.5 MHz higher than the picture RF carrier. The location of these carriers within the channel assignment is shown in the diagram. Frequency allocations have been assigned to 82 TV channels. Details of assignments for typical channels are below.

US TELEVISION CHANNEL FREQUENCIES

CHANNEL		FREQUENCY RANGE	PICTURE CARRIER (MHz)	SOUND CARRIER (MHz)
2	VHF	54–60	55.25	59.75
3	VHF	60–66	61.25	65.75
4	VHF	66–72	67.25	71.75
5	VHF	76–82	77.25	81.75
6	VHF	82–88	83.25	87.75
7	VHF	174–180	175.25	179.75
8	VHF	180–186	181.25	185.75
9	VHF	186–192	187.25	191.75
10	VHF	192–198	193.25	197.75
11	VHF	198–204	199.25	203.75
12	VHF	204–210	205.25	209.75
13	VHF	210–216	211.25	215.75
14–83	UHF	470–890		

From Picture to Video Signal to Picture

The video signal carries information from the scene in the transmitter studio to the face of the TV screen in the receiver. The manner in which this is done will be reviewed very briefly here.

The basic system uses a TV camera tube in the transmitter studio as the transducer to convert light-signal inputs to electrical signals and the picture tube in the TV receiver as the transducer to convert electrical signals back into light output. In the TV camera, a lens system focuses an image of the scene on the face of the camera tube. This face has a light-sensitive surface made up of minute areas that have photoelectric properties. The magnitude of the charge on each of these minute areas is proportional to the amount of light at that part of the image.

HOW PICTURE SIGNALS ARE PRODUCED BY THE TV CAMERA

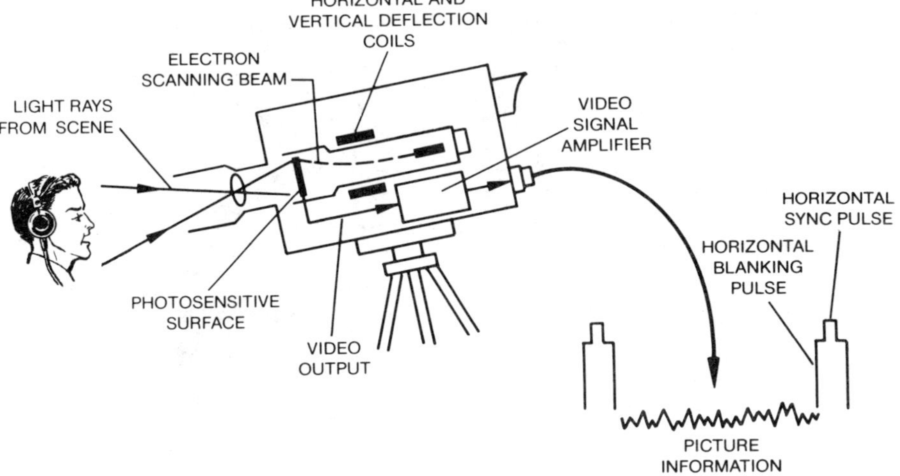

An electron gun in the camera tube focuses a beam of electrons directed toward the light-sensitive surface. Electric currents flow through coils of wire (deflection coils) placed around the neck of the camera tube and form magnetic fields that deflect the electron beam during its travel to the light-sensitive surface. The beam is scanned sequentially—vertically and horizontally—in a series of 525 horizontal lines, one above the other, to cover all parts of the image on the light-sensitive surface. The *horizontal* scan is at uniform speed from left to right with a much faster return (*retrace* or *flyback*) to the left. The *vertical* scan is at uniform speed from top to bottom with a much faster retrace to the top of the screen.

The light-sensitive element produces an electrical output signal when the electron beam strikes it, as described in Volume 3, and this signal is connected to the input of a video amplifier. The magnitude of the voltage produced is proportional to the light intensity at that point on the photosensitive surface. This video information signal is mixed with synchronization and blanking information, and is used to amplitude modulate the RF picture carrier from transmission to the receiver.

From Picture to Video Signal to Picture (continued)

At the receiver, the amplitude-modulated RF carrier is converted back to a video signal. This signal is used to drive the grid in an electron gun in the receiver picture tube (*cathode ray tube*, or *CRT*). The electron gun generates a fine beam of electrons that is focused onto a fluorescent screen at the face of the tube. The beam causes the spot on the screen to flow with a brightness that is proportional to the beam intensity. When a *peak* in the video signal, corresponding to a *bright* spot on the face of the camera tube, is applied to the grid of the electron gun, *more* electrons flow into the beam, causing a brighter spot to appear. When a *lower* point in the video signal, corresponding to a *darker* spot on the face of the camera tube, is applied to the grid of the electron gun, *fewer* electrons enter the beam, causing the spot to be less bright.

CONVERTING THE VIDEO SIGNAL BACK TO A PICTURE

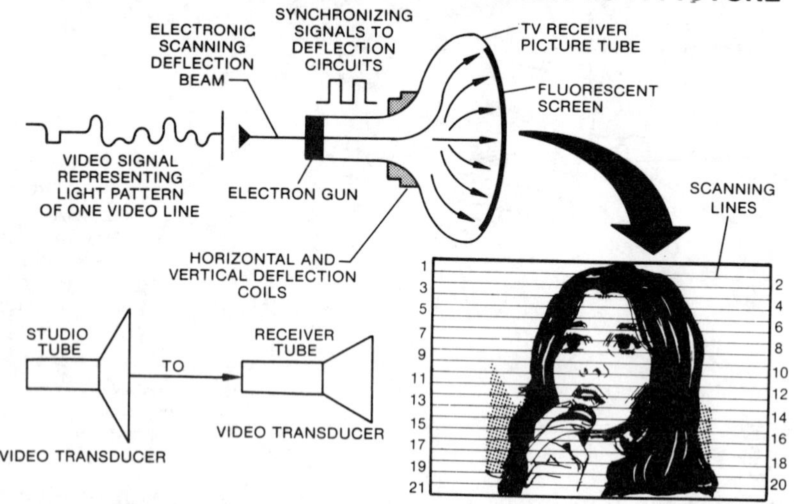

While the *beam intensity* is *varied* in accordance with the video signal, the beam is also *scanned* across the face of the CRT in a *pattern* duplicating the motion of the beam scanning the face of the TV camera tube. The two beams are kept in step (synchronized) by *synchronizing signals*, which are separated from the video signal to *lock* the scanning in the CRT beam to that of the camera tube. The actual scanning is usually accomplished by magnetic fields generated by vertical- and horizontal-deflection coils around the neck of the picture tube through which electric currents flow, duplicating those flowing through coils in the camera tube.

Thus, the *intensity* of the electron beam in the CRT *duplicates* the intensity of the current at the output electrode of the camera tube; also, the *relative position* of the moving spot on the CRT screen *matches* the position of the spot in the camera tube. This *duplication of spot position and intensity* produces an image on the picture-tube screen which duplicates the image produced on the face of the camera tube. The *image* formed on the CRT consists of a *pattern of varying-intensity light* laid down *sequentially*.

Interlaced Scanning

Since the image is formed sequentially, some means must be found to keep the *beginning* of the image *in view* while the rest of the image is being formed. The fluorescent screen and the human eye work together by *retaining* the spot images long enough so that a *complete image* is seen. The process is repeated fast enough so that *no flicker* is observed. Black-and-white picture tubes use a P4 phosphor as the fluorescent material. This is actually a mixture of phosphors to produce white light when struck by the electron beam. The P4 phosphor has a medium persistence so that the image from one scan begins to fade by the time the next scan begins.

If the scanning process that produces the TV image is slowed down, the image appears to flicker. What happens is that there is a *noticeable* delay between the formation of one image and the next. Not only is there a noticeable blank time between images, but the top of the image could fade before the moving spot reaches the bottom of the screen. To avoid this effect in modern TV, 30 complete 525-line pictures are scanned every second. Even though this is a fast rate, flicker can occur, so it is reduced even more by *interlaced scanning.*

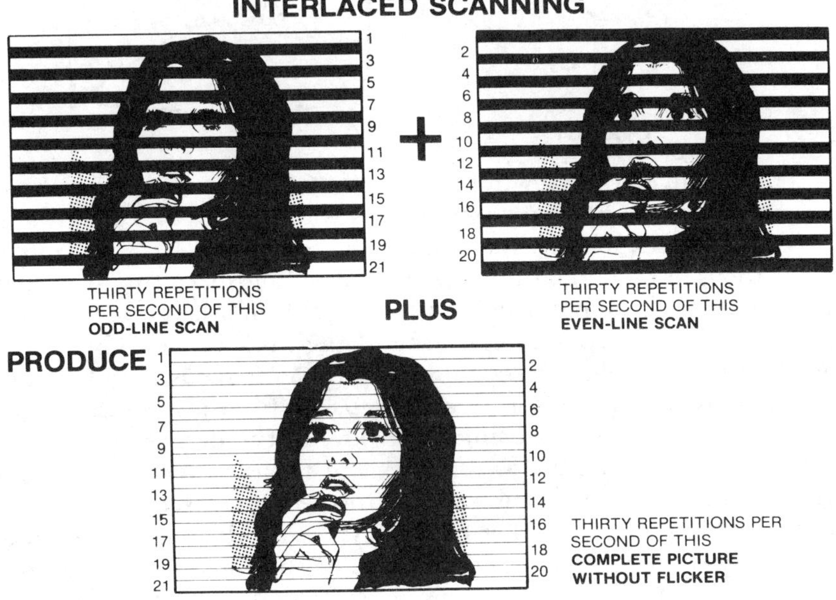

In interlaced scanning, all the *odd-numbered* lines in a 525-line picture (lines 1, 3, 5, etc.) are scanned in 1/60 second. Thus, only 1/60 second elapses between the formation of the top and bottom lines, and there is no noticeable flicker in that time. Then the process repeats for the *even-numbered* lines (lines 2, 4, 6, etc.) in the next 1/60 second. The complete 525-line image formed every 1/30 second is known as a *picture* or *frame.* The images formed every 1/60 second by all the odd-line or even-line scans are known as *fields.*

The Picture Tube (CRT)

The TV picture tube (CRT) is a glass envelope or bulb having the approximate shape shown in the diagram. At the viewing end, the tube is approximately rectangular and the width of the rectangle is 4/3 of the height. These proportions (called the *aspect ratio*), correspond to those of the image scanned at the TV camera in the studio. Early CRTs were circular rather than rectangular. However, the 3/4 rectangular aspect ratio was still used, which resulted in wasted CRT faceplate area. Circular CRTs are still used for many non-TV applications. As its opposite end, the tube narrows down to a cylindrical neck usually between 1 and 2 inches (2.54 and 5.08 cm) in diameter. The neck terminates in a base with a number of pins to which electrical connections are made to the various elements of the electron gun inside. Connection to the high-voltage final anode is made to a terminal on the glass envelope.

THE TWENTY-THREE INCH PICTURE TUBE

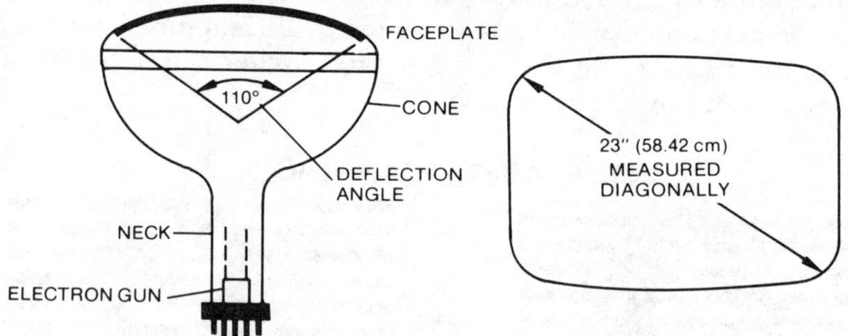

The inside of the viewing end of the CRT is coated with the fluorescent material. The beam of electrons is formed by the electron gun located in the neck of the tube close to the base. The electrons from the gun are focused by magnetic or electrostatic fields to a fine spot that is accelerated by a high potential toward the screen. Magnetic focusing is accomplished by a separate coil around the neck of the tube through which an adjustable dc current flows. Electrostatic focus, which is almost universally used in newer sets, is accomplished by adjusting the voltages of one of the elements (focusing electrode) in the gun assembly. Thus, the beam travels straight down the axis of the tube toward the center of the screen and a bright, fine spot is formed at the screen.

Magnetic coils (or sometimes electrostatic deflection electrodes) deflect the electron beam across the face of the CRT, so that an image can be formed. To do this the beam must be deflected from its straight path through a deflection angle determined by the geometry of the tube. Early tubes were much longer than the one shown and had much smaller deflection angles because efficient deflection circuits did not exist. Public demands for large screens and sets with short front-to-back dimensions have led to present-day tubes with deflection angles of 110 degrees or greater, resulting in much shorter tubes for a given screen size. Tube *size* is always measured diagonally. Popular sizes are available in about 1-inch (2.54-cm) increments between 5 and 23 inches (12.7 and 58.42 cm).

The Picture Tube (CRT) (continued)

Operation of the picture tube is based on vacuum-tube principles. In a vacuum, heated oxide surfaces emit large quantities of electrons that accumulate in a cloud. If an electrode near the oxide surface is made positive with respect to the oxide surface, the electrons travel toward it. The surface that emits the electrons is known as the *cathode*, and the positive electrode is known as the *anode*.

In a picture tube (CRT) there is an electron gun assembly similar to that shown in the diagram. A cylindrical cathode with an oxide-coated end emits electrons when heated. Inside the cathode cylinder is a heater element. When a current is passed through the heater (or *filament*) it raises the temperature of the cathode, and electrons are emitted.

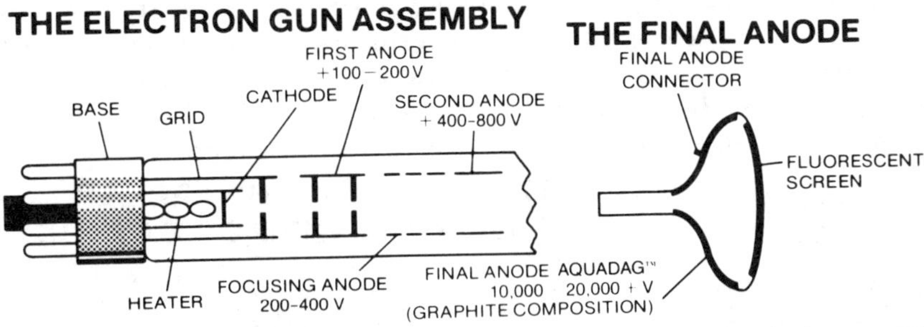

Picture tubes typically contain three anodes, as shown here. Electrostatically focused tubes may have an additional focusing electrode between anodes 1 and 2. The first anode is charged with a positive voltage of about 100 to 200 volts with respect to the cathode and is shaped like a drum with a small round hole punched in the center of each end. The second anode has a positive voltage of about 400 to 800 volts and is shaped like a hollow cylinder. The final anode is a metallic layer plated over the conical portion of the inside surface of the CRT envelope and has a positive voltage from about 10,000 to 20,000 volts (or more) applied. Connection to this anode is usually made via a button on the wall of the CRT. The glass envelope usually has a graphite-composition conducting layer that is grounded coating the outside of the conical portion. The capacitor formed by this coating and the final anode is used to provide filtering of the high voltage for the final anode.

When the electrons are attracted by the positive charge on the first anode, they pass through the small holes because of the much greater attraction of the fields from the second and final anodes. Passing through those small holes forms the electrons into a narrow beam, and the cylindrical shape of the second anode produces an electric field that maintains that beam shape. For electrostatically focused tubes a special additional anode, typically located between anodes 1 and 2, is used as a focusing electrode, and the voltage is adjustable to provide electrostatic focus. In other cases, beam focusing is magnetic. In this case, the beam is confined (or focused) by a concentric, steady magnetic field. Under the influence of the powerful electric field of the final anode, the beam accelerates to the screen and forms a bright spot of light at the center.

The Picture Tube (CRT) (continued)

The grid (control electrode) is a metal disk with a small central hole. It is located between the cathode and the first anode, but relatively close to the cathode. The grid has a negative voltage with respect to the cathode, and, because it is close to the cathode, this negative voltage has a *greater effect* on the electron flow than the anodes. Since these voltages are relative, the grid can be made relatively negative by applying a negative voltage to the grid with the cathode grounded or a positive voltage to the cathode with the grid grounded.

If the grid is made 50–100 volts more negative than the cathode, all the electrons are repelled back to the cathode, and no electrons can enter the beam. Thus, no spot is produced on the screen. If the negative voltage is reduced slightly, the strong positive field of the first anode attracts a few electrons through the grid opening. A beam containing only a very few electrons is formed, and this beam is accelerated toward the screen by the high positive voltages on the anodes. When this beam strikes the screen, a weak spot is formed on the screen.

THE VIDEO CAN BE APPLIED TO EITHER THE GRID OR CATHODE

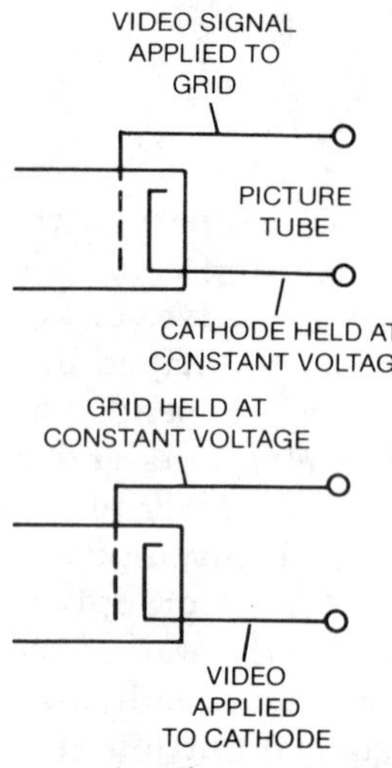

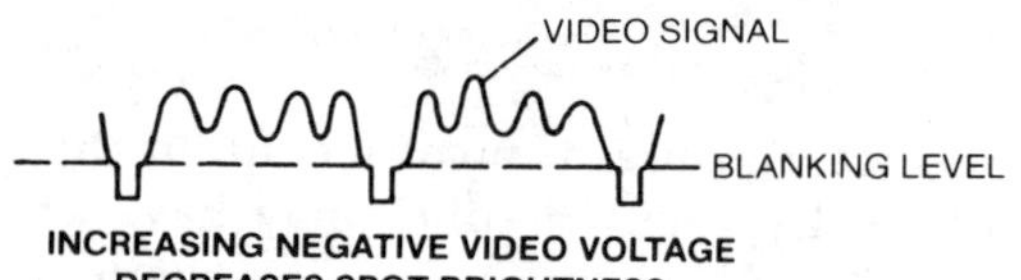

INCREASING NEGATIVE VIDEO VOLTAGE DECREASES SPOT BRIGHTNESS

INCREASING POSITIVE VIDEO VOLTAGE DECREASES SPOT BRIGHTNESS

As the negative voltage on the grid is reduced more and more, increasingly larger numbers of electrons enter the beam, and the spot becomes brighter and brighter. For a typical black-and-white TV picture tube, a voltage change of only about 30 volts is sufficient to change the spot intensity from total black to maximum useful brightness.

In the TV receiver, the video signal is applied between the cathode and grid with the polarity such that an increasing signal level—toward the blanking level—results in decreasing brightness. Thus, a black-level signal is sufficient to cut off the flow of electrons completely. The amplitude variations in the video signal cause corresponding brightness variations in the spot on the screen via the grid. When the spot is moved across the screen in horizontal and vertical lines (scanned), these brightness variations produce the picture on the CRT screen. As you can see from the illustration, the video signal can be applied to either the grid or the cathode, as long as the correct polarity is used.

Scanning the Electron Beam

A TV frame consists of 525 horizontal lines scanned sequentially from top to bottom of the frame in 1/30 second. Each frame consists of two interlaced fields, each containing 262½ horizontal lines scanned in 1/60 second. Thus, to form each such frame requires that the electron beam be scanned 262½ times in a horizontal direction for each scan in a vertical direction.

In *Basic Electricity*, you learned that a circular magnetic field is formed around a wire when an electric current flows through the wire. The direction of the magnetic field is determined by the direction of current flow, according to the left-hand rule. You also learned that if a current-carrying wire is placed between the poles of a magnet, a force will act on the wire. The direction of motion will be to the left or right, according to the current direction and magnetic polarity, as described in the three-finger rule. This motion is the basis of electric motor operation.

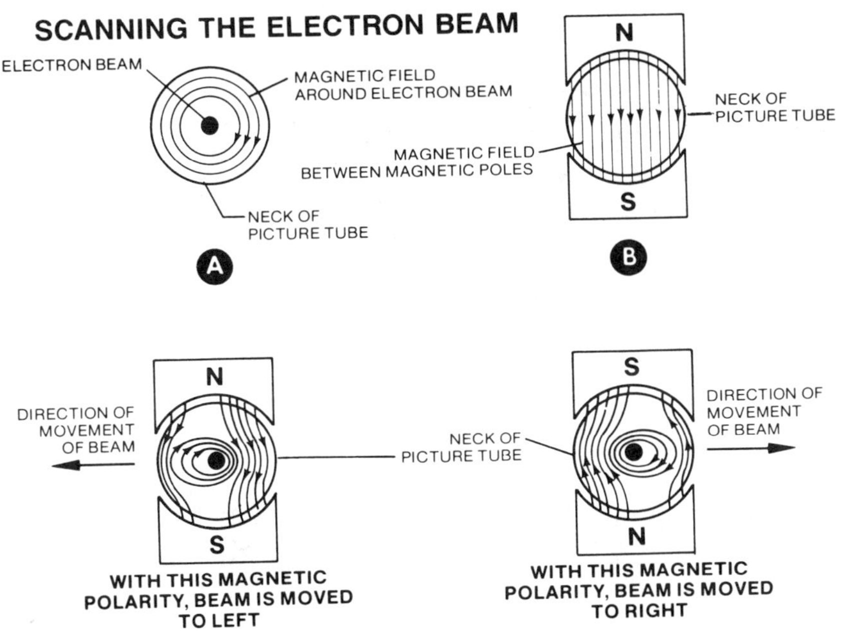

This motion is also the basis for electron-beam deflection in a picture tube. The electron beam in its path from the gun to the screen has the same properties as current or electron flow in a wire—each consists of electrons in motion or an electron stream. Even though this stream is not confined by a wire, in a CRT it forms a magnetic field around itself in the same manner as if the wire were there.

Now, if magnetic poles are placed above and below the beam, the beam is deflected by the force generated by the interaction of the two magnetic fields in the same manner as the current-carrying wire. Since the beam has very little mass, it can be deflected very rapidly. The direction of the deflection is reversed by reversing the polarity of the magnetic poles.

Scanning the Electron Beam (continued)

In a television receiver, the magnetic poles are formed by coils of wire—the deflection coils—close to the neck of the tube, as shown in the diagram. Now the strength and polarity of the magnetic field can be changed as desired by changing the magnitude and direction of the current flow through the coils.

A pair of coils above and below the electron beam (H1 and H2 in the diagram) produces the desired horizontal deflection. Another pair of coils at the left and right of the beam (V1 and V2 in the diagram) produces the desired vertical deflection. The composite set of coils is usually called the *deflection yoke*. These coils are fitted around the neck of the CRT at the point where the neck flares into the cone.

DEFLECTION COILS

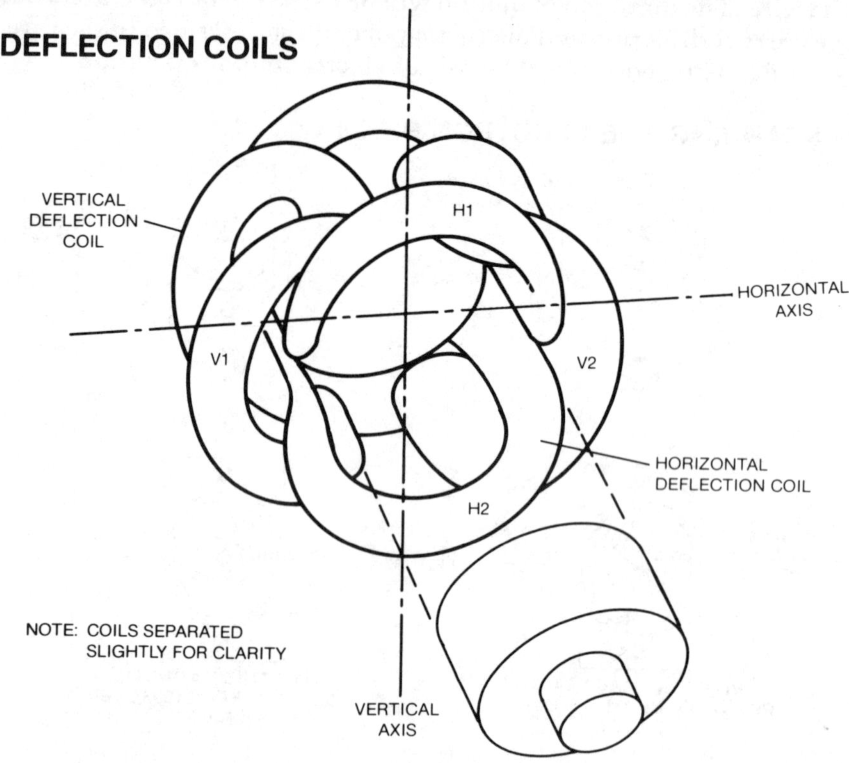

To produce one of the interlaced frames of a complete picture requires simultaneous vertical and horizontal motion of the electron beam by the deflection coils. To sweep the beam uniformly from the extreme left to the extreme right with a very fast retrace to the left requires a sawtooth magnetic field that makes 262½ cycles in 1/60 second (or 262½ × 60 = 15,750/second). During that same period, another sawtooth magnetic field produces a uniform deflection from top to bottom with a very fast retrace to the top every 1/60 second or 60/second. Since the scan is interlaced, two 1/60 second fields make up one frame at a 30-Hz rate.

Scanning the Electron Beam (continued)

To obtain a sawtooth of current through an inductance requires a constant applied voltage since, as you may remember, $\Delta_I/\Delta_T = E/L$. A sawtooth is defined as having a constant rate of rise, or Δ_X/Δ_T is a constant where Δ_X is any quantity. Thus for a sawtooth magnetic field, the current must vary linearly, or Δ_I/Δ_T is a constant. Therefore E/L is a constant, which means that E is a constant since L (the inductance) does not change. Therefore, rectangular waves of voltage across the coil produce a sawtooth of current. During retrace, the current is abruptly interrupted, as shown. Thus, the magnetic field collapses, bringing the beam rapidly to the beginning of the scanned line or field. Because the deflection yoke coils have resistance, a small sawtooth component is added to produce a linear magnetic flux, as shown. The sawtooth component compensates for the voltage drop across the resistance as the current increases. Thus, the voltage across the inductance remains constant.

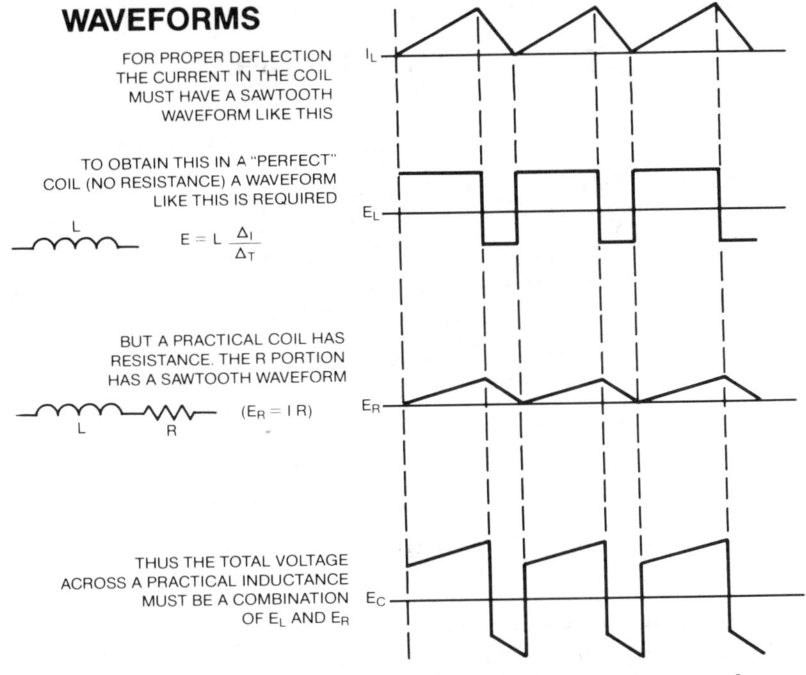

The simultaneous horizontal and vertical scanning produces the *raster* of horizontal lines, one above the other. When you switch to a channel that is not in use, you can see the array of horizontal lines, all of uniform density, that make up the raster. You know that the electron beam is deflected from left to right by a sawtooth wave. Starting at the extreme left, the spot on the screen is moved at uniform speed to the extreme right with a very fast retrace or *flyback*. If you look closely, you may see the retrace on the TV set if it is tuned to an unused channel. The TV video signal has a blanking pulse as part of the synchronization that applies a high negative voltage to the grid of the CRT during the retrace period to keep retrace lines from appearing on the screen.

Scanning the Electron Beam (continued)

You know that the scan is interlaced and that an odd number of horizontal scan lines (525) makes up a complete frame. As described earlier, interlaced scanning is used to prevent flicker. This is done by scanning a field consisting of all the odd horizontal lines in a picture (1/60 second), and then scanning a field consisting of all the even horizontal lines (1/60 second) to make up the complete field (1/30 second). Because a picture is divided into two fields, each field consists of exactly 262½ horizontal lines. The scan for line 1 starts at the extreme upper left. The last odd-line scan thus ends with a half-line scan, and the next even-line frame must start with only half a horizontal scan line. The odd/even line scans are synchronized by the synchronizing signals present during the blanking period. As you learned in Volume 3, the odd/even line scans are automatically produced by the synchronizing signal for the vertical sweep so that an interlaced scan is produced from the transmitted signal, not as a result of any special circuits in the receiver.

INTERLACED SCANNING

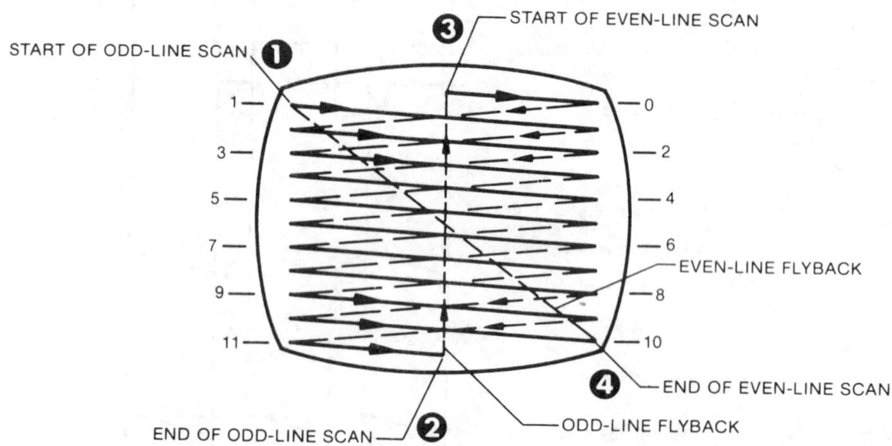

As you know, horizontal synchronization is maintained during the vertical blanking period. The vertical-sync pulse is serrated by narrow notches that occur at the horizontal-sweep rate to maintain horizontal sync. The region of the vertical-sync pulse preceding and immediately following the vertical-sync pulse contains equalizing pulses that occur at twice the horizontal-line rate. Since the horizontal-sweep circuits are sensitive to only the pulses that occur in the vicinity of the natural period of the horizontal-line oscillator, horizontal sync continues as before during the equalizing-pulse period; on alternate sweeps, however, the vertical sync starts earlier by one equalizing pulse (half line). To keep the sweep lines constant, the equalizing pulses after the vertical sync contain one additional pulse. Thus, effectively, the vertical-sync pulse is moved half a horizontal line every other sweep to provide for interlacing.

The Video Signal

Now what you need to learn more about at this time is *how* the picture on the receiver screen is *kept stable* by synchronizing the scanning in the CRT with the scanning in the camera tube. These signals were discussed in detail in Volume 3. In the figure below, details are shown only for five even-numbered scans over lines 200 through 208. The white level is shown as the lowest part of the scan and the black level as the highest. It is important to note that a black (blacker than black) blanking pulse is inserted at the end of each scan. Its purpose is to blank the retrace of the electron beam by turning the beam off.

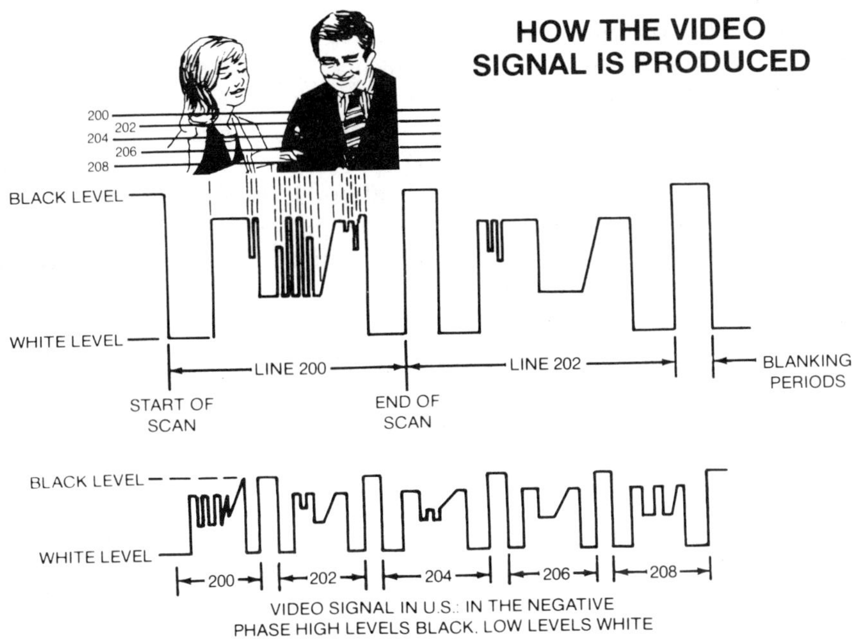

VIDEO SIGNAL IN U.S.: IN THE NEGATIVE PHASE HIGH LEVELS BLACK, LOW LEVELS WHITE

As an example, suppose you trace out the details of scan number 200. At the extreme left is the bright background of the scene, which produces a corresponding width of maximum white signal (minimum carrier level). Then the scan passes to the very dark hair of the scene and produces a matching width of maximum black signal (high carrier level). The left side of the cheek again produces a maximum white signal, but the level then varies rapidly over gray levels as the scan passes over the varying shades of gray of the nostrils and right cheek. Maximum black is then produced by the hair at the right side of the image except for three small highlights. At the extreme right of the image is the maximum white of the background. At the end of each line is a slightly blacker-than-black-level signal (the *blanking pulse*) that is introduced into the video to turn the CRT beam *off* during retrace. Thus *horizontal-blanking pulses* are added to the video signal, starting shortly before retrace starts and ending slightly after the new line scan begins. A similar *vertical-blanking pulse* is inserted for the vertical retrace.

Sync Pulses

The horizontal-sync pulses occur during the horizontal-blanking period and represent 100% (maximum) modulation of the picture carrier. A horizontal-sync pulse occurs on each horizontal-blanking pulse and is used to initiate the horizontal-sweep retrace and beginning of a new sweep line, as shown.

During the vertical retrace, it is necessary to have a *blanking pulse* so that the vertical retrace will not appear on the screen. It is also necessary to have *vertical-sychronizing pulses* to start the vertical sweep at precisely the right moment. The operation for vertical blanking and synchronizing is similar to that for the horizontal, with the exceptions noted below. The illustration on the following page shows the structure of the transmitted signal during vertical blanking.

Vertical retrace is slower than horizontal because the vertical sweep is slower; therefore the blanking period must be longer. Typically, the vertical blanking of each field requires the same amount of time as about 18 horizontal lines, so that actually only about 490 active horizontal lines are seen in a complete frame. Vertical-sync pulses are much wider than horizontal-sync pulses, and they are distinguished from horizontal-sync pulses on this basis, as you will learn later.

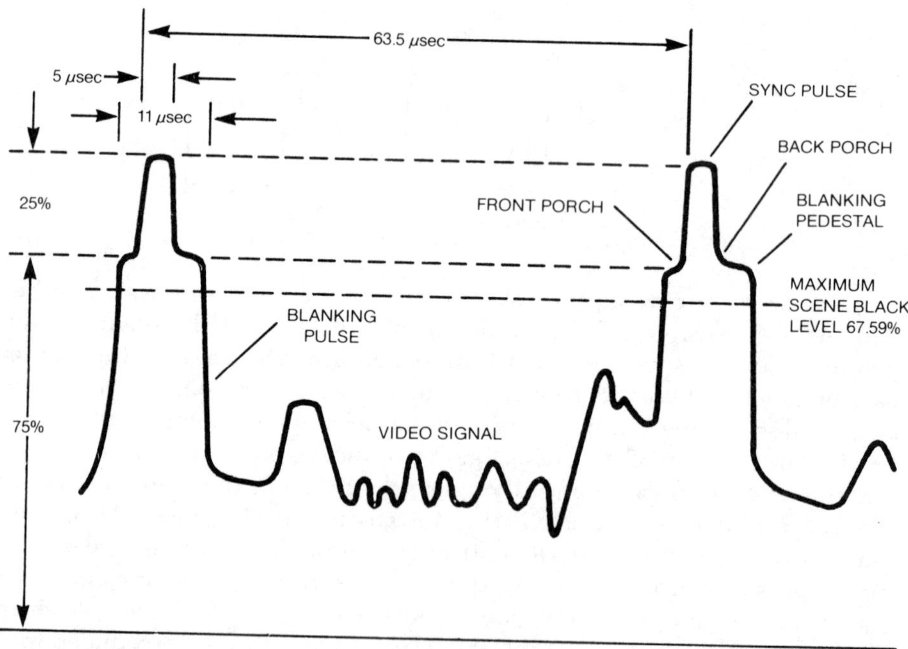

Sync Pulses (continued)

Provision must be made to continue supplying horizontal-sync pulses during the vertical-sync period so that the horizontal scan will not be out of sync when it must begin to scan the top line of the next field. Finally, because a half frame is made up of 262½ horizontal scans, provision must be made for synchronizing the half of a horizontal line that occurs alternately at the top and bottom of every other field. All these requirements are met by having a vertical-blanking and synchronizing pulse such as that shown here.

When vertical blanking occurs, a series of 5 narrow equalizing pulses are generated at twice the line rate (5 pulses in 2½ horizontal line intervals during field 1 and 6 equalizing pulses for field 2). Every other one of the equalizing pulses is used to keep the horizontal oscillator in sync. The vertical-sync pulse consists of 6 or more pulses at twice the horizontal rate, except that these pulses are very wide, as shown. The vertical-sync pulse is followed by 6 more equalizing pulses for field 1, but only 5 (2½ lines) equalizing pulses for field 2. Thus, the second field starts half a line later, yielding an interlaced scan between field 1 and 2.

In the receiver, the sync separator distinguishes between the narrow horizontal-sync pulses and the wide series of 6 pulses that make up the vertical-sync pulse.

FCC STANDARD TELEVISION VERTICAL SYNCHRONIZING SIGNAL

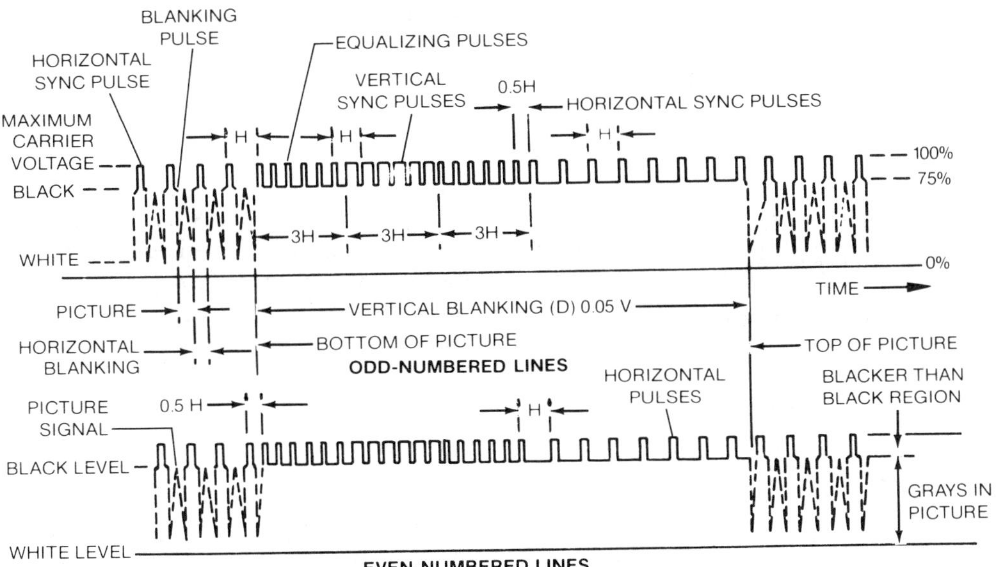

Review of Television Reception

1. **TV CONVEYS VISUAL AND AURAL INFORMATION** either through the air via radio waves, or through conductors in *closed circuit TV* (CCTV) or *cable TV* (CATV).

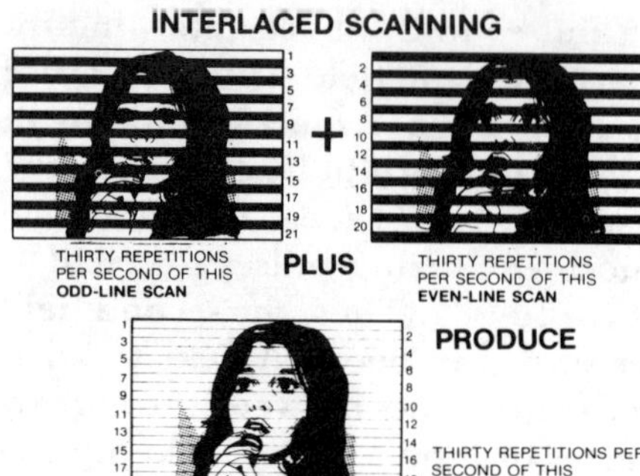

2. **INTERLACED SCANNING** covers the image, both horizontally and vertically, in two steps. Each step is called an *interlaced field*.

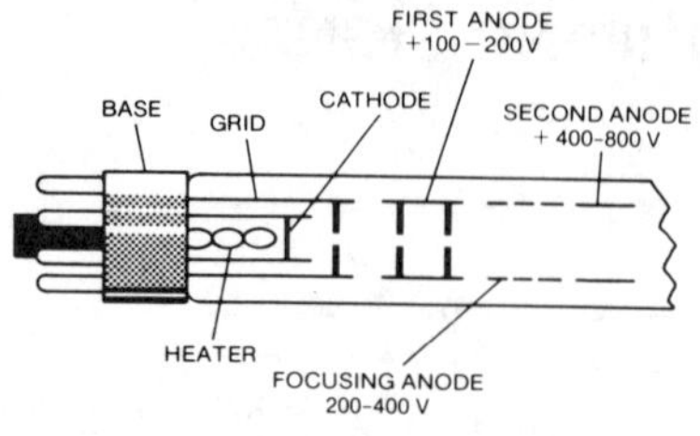

3. **THE PICTURE TUBE** generates a beam of electrons directed at a fluorescent screen. The beam originates in an *electron gun*.

4. **THE VIDEO SIGNAL** results from different voltage outputs as the camera beam scans light and dark elements. In the receiver, the picture tube beam intensity is varied to build the image.

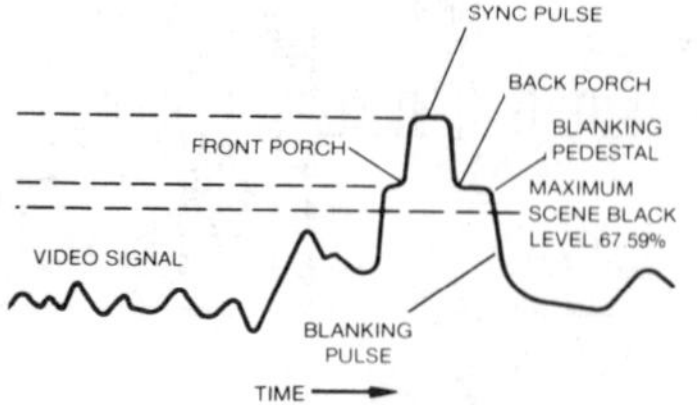

5. **SYNC PULSES** reach a 100 percent modulation level. Each horizontal pulse is mounted on a *blanking* pulse. Vertical pulses are formed of groups of wider horizontal pulses.

Self-Test—Review Questions

1. What are CATV, CCTV? How are they different from commercial TV?
2. Describe briefly the essential elements of a TV system.
3. What is interlaced scanning?
4. Why is interlaced scanning used?
5. Draw a sketch of a CRT. Show how it operates to produce a TV image.
6. How is the electron beam in CRT scanned?
7. Draw a sketch of a TV signal.
8. What are the frame field and line rates for commercial TV in the United States?
9. Sketch and describe the black (blanking level) and the horizontal- and vertical-sync pulse characteristics. What do they do?

Learning Objectives—Next Section

Overview—Now that we have reviewed the fundamentals of the TV system, we next proceed to a study of the details of TV receivers, starting with the black and white type.

VHF-UHF TUNERS

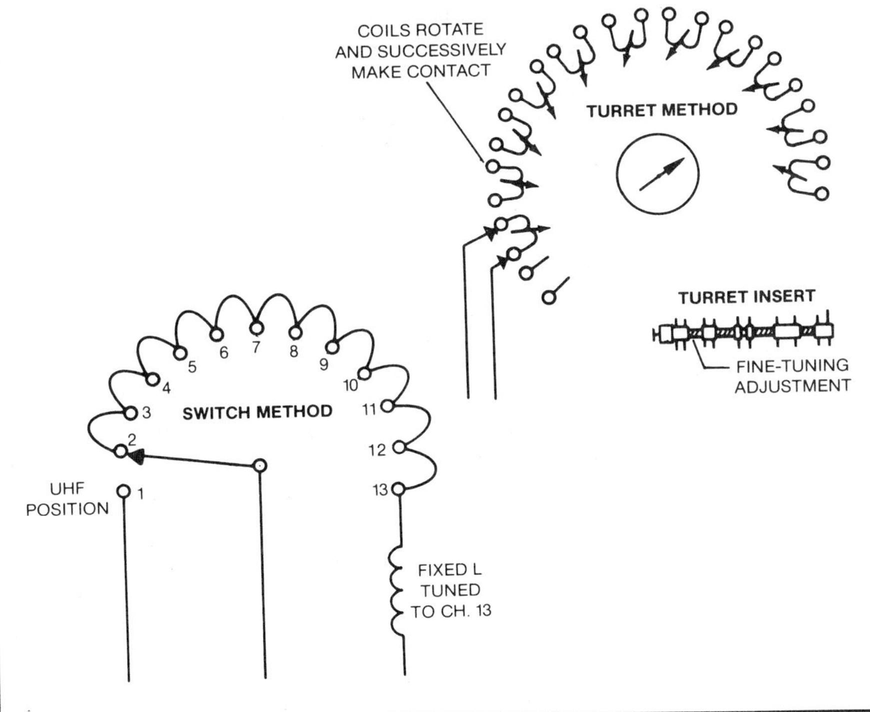

Typical TV Receiver—Block Diagram

Now that you are familiar with the signals used for transmission of television information, you are ready to learn the details of how black-and-white TV receivers operate.

The following page is an overall block diagram of a typical black-and-white TV receiver. The VHF and UHF tuners are fed from their respective antennas and include the RF amplifiers, mixers, and local oscillators. The UHF tuner output is fed to the VHF tuner so that the RF amplifier and mixer can be used as a preamplifier to provide additional gain and to provide double conversion to improve the image rejection in the UHF range. The IF amplifier usually consists of several stages with the necessary tuned circuits to provide selectivity and to provide the special shaping of the passband required to receive the vestigial-sideband transmission. The IF signal is demodulated to video by the video detector. This detector also produces as the output a 4.5-MHz sound carrier that is fed to the FM sound circuits via the video amplifier. These sound circuits include amplifiers, the FM demodulator, and the audio amplifiers to drive the loudspeaker.

The video signal from the detector is amplified for application to the grid/cathode of the CRT to control the beam intensity in accordance with the video data. An AGC circuit is provided via the ouput video to control the gain of the IF amplifier and the tuner RF amplifier.

An output from the video amplifier is also used to drive the sync separator, which develops horizontal- and vertical-sync pulses for the horizontal- and vertical-sweep generators. These signals provide the information necessary to keep the sweeps in the receiver synchronized to the transmitted signal. The sweep outputs drive the deflection yoke so that the beam of the CRT moves synchronously with the scanning beam in the TV camera, while the video signal modulates the beam intensity in accordance with the brightness of each point on the picture.

The high voltage necessary for the final anode of the CRT is obtained from the horizontal-sweep circuit during retrace. A low-voltage power supply provides the low voltages necessary for operation of the receiver circuits.

As before, we will use an actual system, a portable black-and-white TV receiver, as a learning system not only to help understand the individual circuit functions, but also to understand how the TV receiver functions as a system. Here, the learning system will be the model 12HB1 portable television receiver manufactured by the Zenith Radio Corporation. The 12HB1 has 26 transistors, 27 diodes, and a single IC. It covers VHF Channels 2 through 13 with memory fine tuning and UHF Channels 14 through 83. Some modern black-and-white TV sets use more ICs, but this set has been chosen because it clearly illustrates the necessary principles with easily understood circuits.

BLOCK DIAGRAM OF A TYPICAL BLACK-AND-WHITE TV RECEIVER

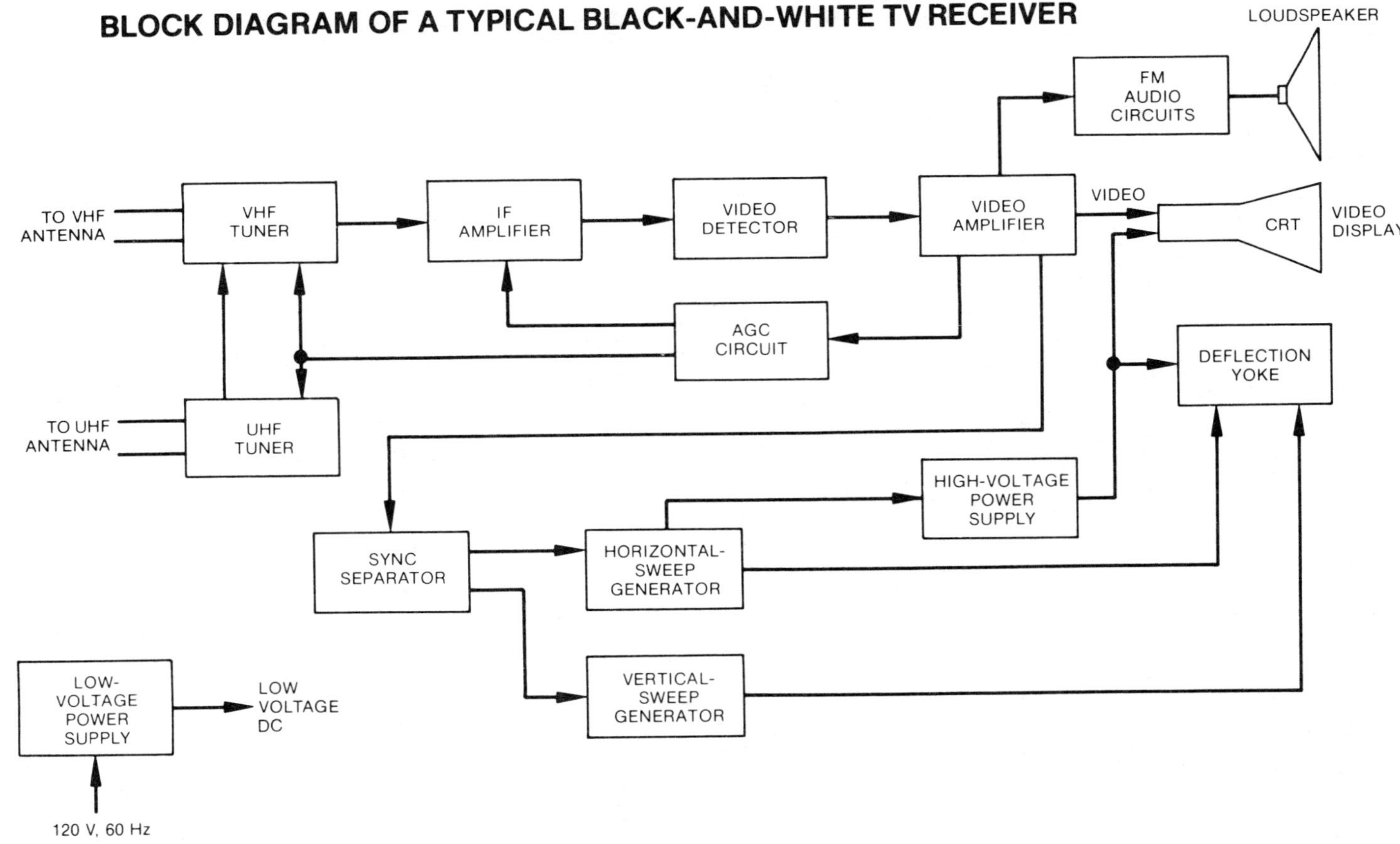

Receiver Section—Block Diagram

The block diagram of the receiver circuits that produce the video, sound, and synchronizing-signal complex for the other sections of the receiver is shown on the following page. Note that the blocks are marked with the circuit-identification numbers (Q1, Q2, etc.) to relate this diagram to the detailed schematics to be described shortly. The receiver must operate on either VHF or UHF channels, and separate tuners are used to process these signals most efficiently. The VHF tuner accepts the input from its own antenna and contains an RF amplifier stage (Q1), mixer (Q2), and local oscillator (Q3). The UHF tuner is similar, containing an RF amplifier (Q5) and local oscillator (Q6); the mixer, however, is a crystal diode (CR14). The UHF mixer output from CR14 is at VHF and is fed to the VHF tuner to be converted to the IF of the receiver. Thus, in UHF operation, double conversion is used.

The output of the VHF tuner mixer stage (Q2) is an IF signal containing the video IF carrier at 45.75 MHz and a sound IF carrier at 41.25 MHz. These combined signals are amplified by a three-stage video IF amplifier (Q101, Q102, Q103).

Output from the final (third) video IF is fed into the video detector CR101. The output from this stage consists of the video signals at baseband and the sound FM IF signal at 4.5 MHz. This 4.5-MHz sound IF signal is generated by CR101 acting as a mixer between the video and sound carriers, producing a 4.5-MHz sound IF signal. These video signals are amplified by the video driver Q801 that simultaneously amplifies the video, sync, and sound IF (4.5-MHz) signals. Video outputs to the sync separator and 4.5-MHz sound IF are taken from video amplifier Q801.

The AGC for TV receivers is somewhat different from the AGC systems that you have studied. The reason for this is that the average carrier level changes with scene brightness; therefore the average carrier level cannot be used for AGC. The sync pulses are of constant amplitude (100% modulation) and can be used for AGC. To accomplish this, the sync pulses are gated by Q403, the AGC gate. The AGC gating pulses are derived from the horizontal-sweep circuits so that only the sync pulses are fed to the AGC output (Q402). These pulses are filtered to produce a dc level that is proportional to the peak carrier level (sync level). This signal is applied to the first IF amplifier (Q101). An AGC voltage is also applied to the RF amplifiers (Q1 and Q5) via AGC delay (Q401). The AGC delay prevents the application of AGC voltage to the RF amplifiers except on very strong signals. Thus, full sensitivity is available for weak signals.

The remaining circuits are described next. The output from Q801 drives the final video amplifier Q803 which drives the CRT cathode. As you can see, at least in block-diagram form the receiver section of a TV receiver has the same elements as the receivers you studied earlier.

RECEIVER SECTION BLOCK DIAGRAM

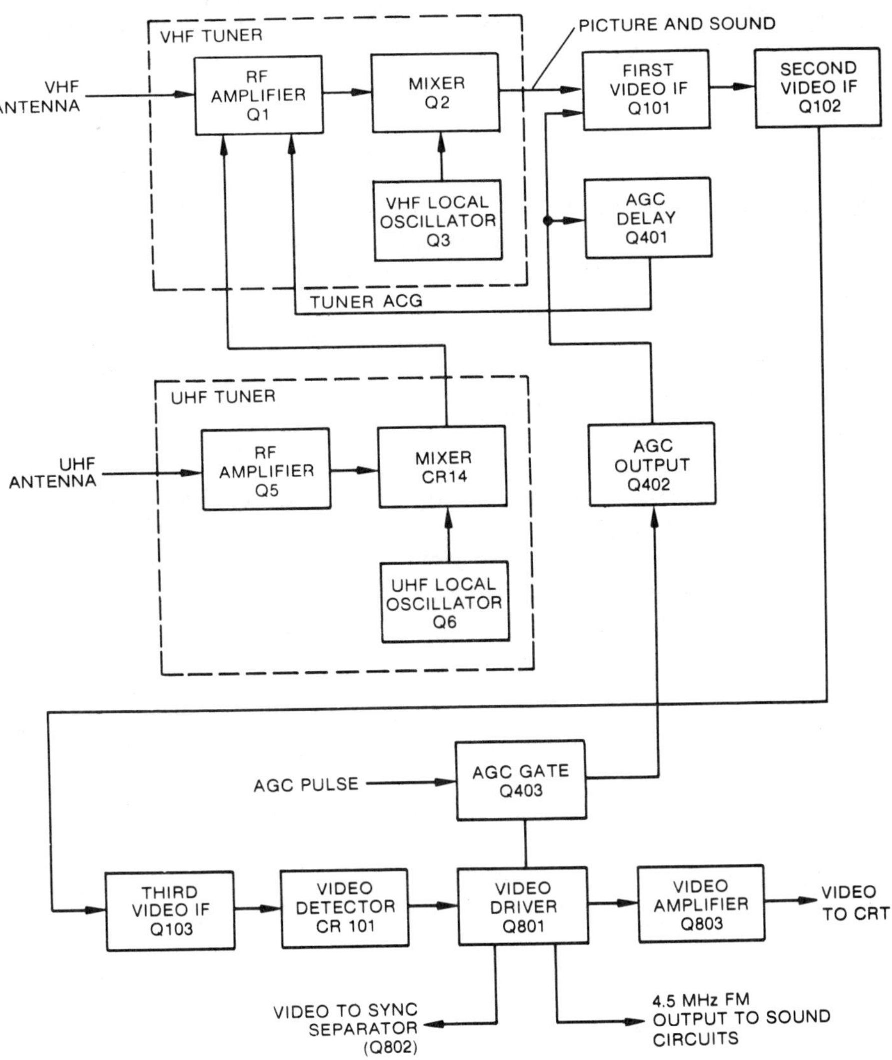

Sweep- and Sound-Circuit Section—Block Diagram

The 4.5-MHz sound IF output from video amplifier Q801 is applied to IC1001. This IC contains an IF amplifier, FM demodulator (ratio detector), and an audio preamplifier. The output from IC1001 (audio) is applied to audio driver Q1001, which drives the complementary-symmetry audio-output amplifiers Q1002 and Q1003. These drive the loudspeaker. The sound IF audio system is similar to the one for FM receivers.

The video output from video amplifier Q801 is also applied to sync separator Q802. The sync-separator (Q802) output consists of sync pulses stripped from the incoming video signal. These sync pulses (horizontal and combined vertical) are further separated (by RC networks to be described later) into horizontal-sync pulses and vertical-sync pulses to control the horizontal and vertical sweeps.

The horizontal-sync pulses are applied to the horizontal APC (automatic phase control) Q501 that controls the frequency of the horizontal oscillator Q503. The horizontal-sync pulses are compared in phase with the retrace pulses from the high-voltage transformer driven by the horizontal output Q504. An error voltage is developed in Q501 that drives the frequency of the horizontal oscillator so that the sync pulse and retrace pulse are locked in phase, thus synchronizing the horizontal oscillator to the horizonal-sync input. The output from the horizontal oscillator is amplified and shaped in horizontal driver Q502 that drives the horizontal output Q504. The output from Q504 drives the deflection yoke and the high-voltage transformer TX503. The current necessary for horizontal retrace is much greater than for the sweep because the retrace time is short. Thus, a high-voltage pulse is generated during retrace of the horizontal sweep. The pulse appears also at the output transformer TX503. This output pulse from TX503 is rectified by CR504 to provide the 12 kV dc high voltage to the CRT final anode. The high-voltage filter capacitor is formed by the final anode inside coating and the bulb outside coating, as described earlier. Because of the high frequency involved (15,750 Hz) and the low current requirements of the final anode, this filtering is sufficient. The AGC pulse is derived (as is the horizontal-retrace pulse for the horizontal APC) from a separate winding on the high-voltage transformer.

The vertical-sync pulses are applied to the vertical oscillator Q601–Q602 to lock it directly to the sync. The output from the oscillator is shaped and amplified by vertical amplifiers Q603 and Q604 that drive the vertical driver Q605. The vertical driver Q605 drives the complementary-symmetry pair of transistors (Q606–Q607) that drives the vertical-deflection coil.

A low-voltage power supply (not shown) is of conventional design, providing an unregulated 24-volt dc signal to the audio circuits with a regulated +22-volt output for all video circuits and a regulated +11 volts for tuner operation. Input is 120-volt ac at 60 Hz.

SWEEP-AND SOUND-CIRCUIT SECTION BLOCK DIAGRAM

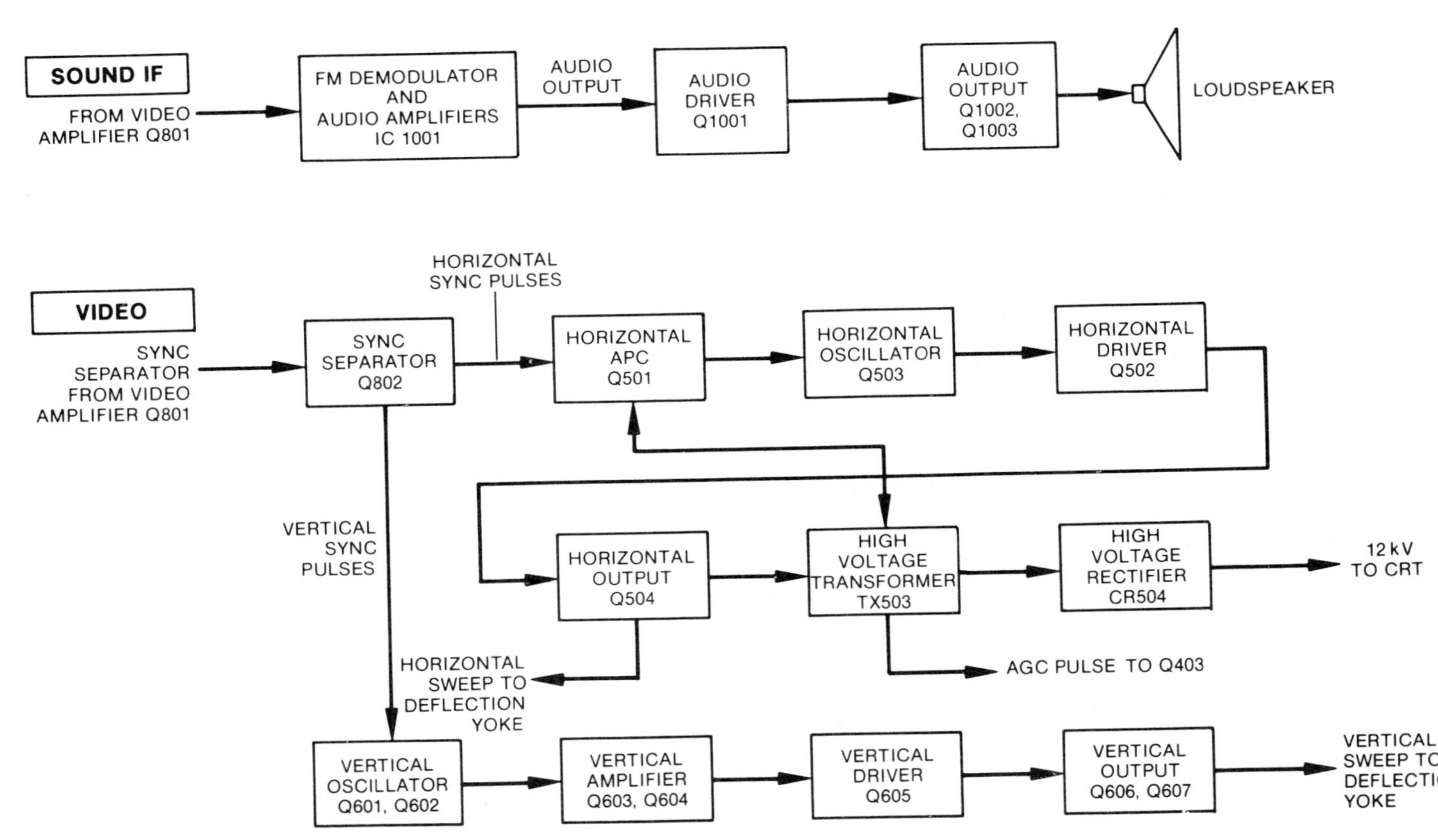

VHF–UHF Tuners

As in most TV receivers, the VHF and UHF tuners are separate units using separate antennas. For the Zenith model 12HB1, the VHF tuner has an input from the UHF tuner output which is selected when the VHF channel selector is in the UHF position. Both the VHF and UHF tuners use an RF amplifier, mixer, and local oscillator. The VHF tuner uses a ganged selector switch to adjust the inductance of the tuning coils for the antenna, RF amplifier, and mixer coils. This is accomplished by a tapped inductor with the proper tap for each channel to provide for tuning to that channel, as shown. The total circuit inductance in creases as the switch is moved from the channel 13 position toward channel 2.

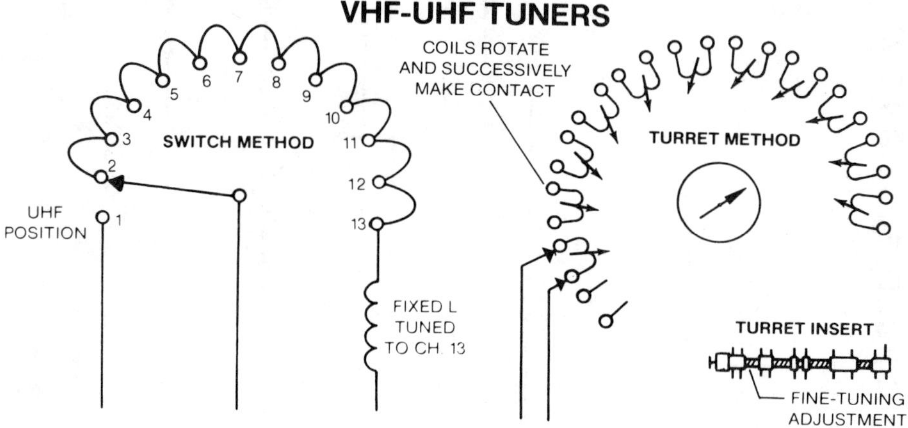

Separate inductors are used for the local oscillator for each channel. A trimmer capacitor is associated with each coil to provide fine-tuning adustment. For a selected channel, this tuning capacitor can be engaged by pushing the fine tuning knob in so that the tuning capacitor is engaged to permit adjustment. When the fine-tuning knob is released, the capacitor for that channel retains its setting so that readjustment is necessary only infrequently, when the local oscillator has drifted off frequency. This so-called *memory tuning* is found on all but the simplest TV receivers.

Some receivers use a turret tuner. For a turret tuner, a separate set of coils is provided for each channel on a rotatable turret. These are switched into the tuner circuits by rotating the turret, bringing the coil contacts for a given channel into contact with a set of fixed contacts on the tuner. Memory tuning is accomplished in the same way as for the tapped-coil tuner, with the trimmer capacitor (or inductor tuning slug) carried on the LO coil end of the coil assembly.

VHF–UHF Tuners (continued)

Many modern, more expensive color TV receivers use a synthesized approach to tuning. In this case, the LO signal is synthesized directly from a phase-locked loop, as described earlier.

The schematic diagram of a typical VHF tuner, such as is used in the Zenith model 12HB1, is shown on the following page. The RF input signal from a balanced 300-ohm line is brought to a low impedance (75 ohms) and converted from a double-ended to a single-ended input by T1. This arrangement is often called a *balun* (*bal*anced to *un*balanced). The LC filter following the balun is a high-pass filter that suppresses all signals below channel 2 to prevent feedthrough of low-frequency interference into the IF amplifier. Transistor Q1 is a common-base RF amplifier with delayed AGC applied to its base. The input circuit for Q1 is tuned by the antenna coil, as shown, in conjunction with the circuit capacitance. Since the circuit bandwidth is directly affected by circuit capacitance, the capacitance is carefully controlled. In addition, the inductive reactance, due to even a short length of wire, is considerable at VHF and UHF. Therefore very short heavy leads are used in the signal path. The gain of Q1 is reduced as the AGC input increases the positive bias on the Q1 base. This increases the current through Q1, which has a 1.2-K resistor in its collector circuit (on the other side of the coil). The increased current increases the voltage drop across the 1.2-K resistor, reducing the collector voltage on Q1 and hence reducing its gain. AGC is applied in this manner to minimize shift in the timing of the input circuit which can result when the gain is directly controlled by controlling base bias. The interstage between Q1 and the mixer (Q2) is tuned by the RF/mixer tapped inductor, stray circuit capacitance, and selector switch, as shown. The mixer (Q2) accepts the signal from the RF amplifier (Q1) and the LO signal from Q3 and produces an output at the IF. The IF output from the mixer is taken from output transformer T3. Transformer T3 is a broadband transformer tuned to the IF. The output is taken at low impedance for coupling with the IF via a coaxial cable. The local oscillator operates on the high side of the signal (45.75 MHz above the picture carrier). Therefore, the picture carrier at IF is at 45.75 MHz and the sound IF is 4.5 MHz lower, or at 41.25 MHz. The local oscillator (Q3) uses a modified Colpitts circuit.

The UHF tuner uses a similar configuration except that tuning capacitors are used with fixed inductance. The UHF tuner has detented positions to adjust the tuning capacitors to the proper channel. In addition, the mixer is a diode rather than a transistor. The output from the UHF tuner is fed to the RF amplifier/mixer. When the VHF tuner is in the UHF position, the VHF local oscillator is not used, and the VHF RF amplifier and mixer are tuned to the IF, providing additional gain before the main IF amplifier.

TYPICAL VHF TUNER

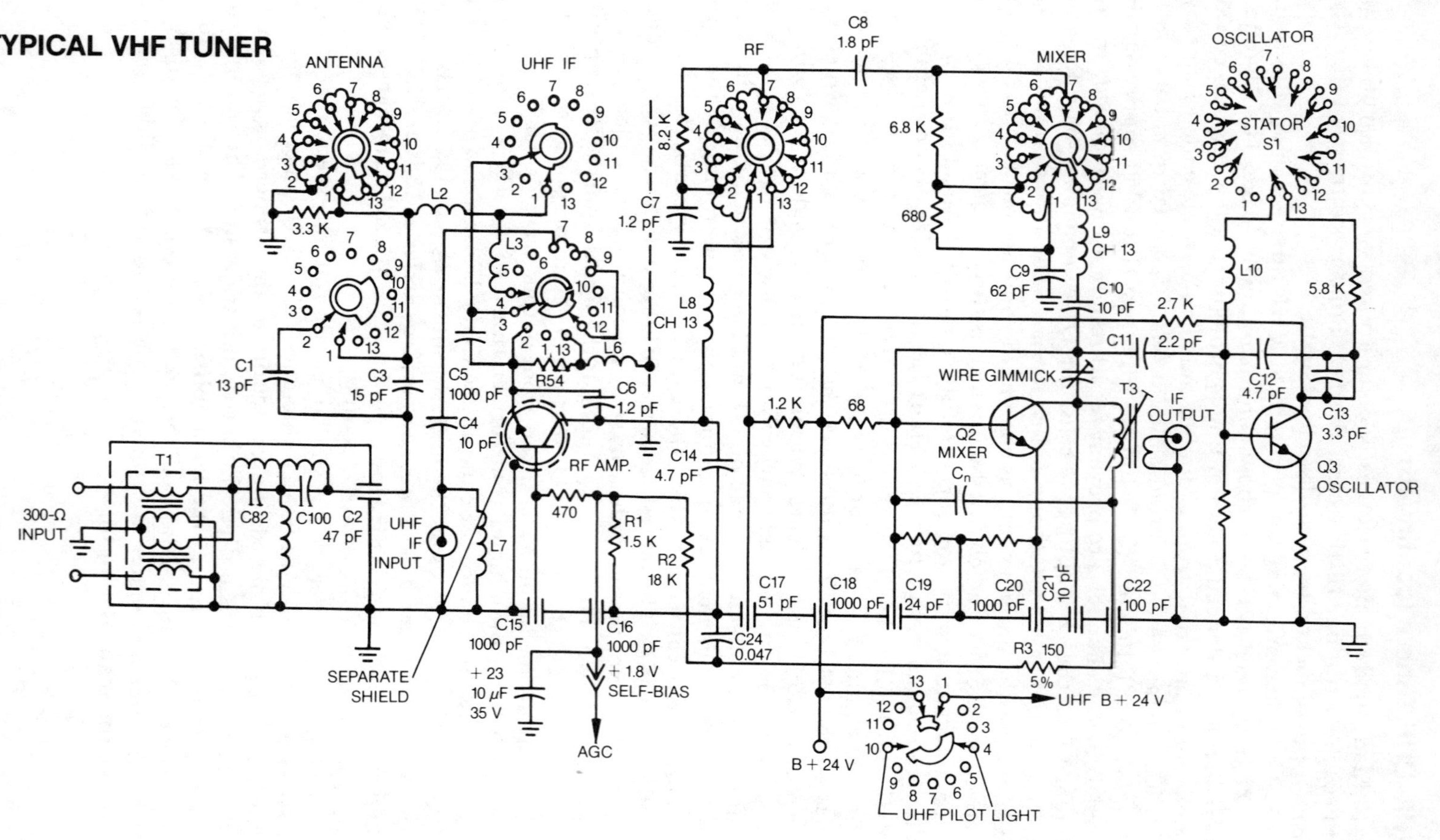

Video IF Amplifier

Most TV tuners have local oscillators tuned to the high side of the desired signals. The nominal IF (picture carrier) for the present-day TV is at 45.75 MHz, and the sound carrier is at 41.25 MHz. The IF amplifier must pass all signals in the band between the picture and sound carriers plus an additional 0.75 MHz for the lower vestigial sideband of the TV picture. Thus, the IF amplifier must be broadband to amplify signals between 41.25 and 46.5 MHz. Because of the vestigial sideband, the lower frequencies in the picture signal (±0.75 MHz) are transmitted double sideband while the higher frequencies (above 0.75 MHz) are transmitted single sideband. This, of course, does not apply to the aural carrier.

Since the sideband power at low video frequencies is twice that of the higher frequencies, special shaping of the IF passband in this region is necessary to equalize the sideband power at all frequencies. This is accomplished by shaping of the IF passband as shown in the illustration. As you can see, the IF amplifier gain is slowly reduced, starting at a point 0.75 MHz below the picture carrier, and is zero at a point 0.75 MHz above the picture carrier. At any given frequency in this region (±0.75 HMz) the sum of the responses on both sides of the carrier is equal to the response of the rest of the lower IF sideband.

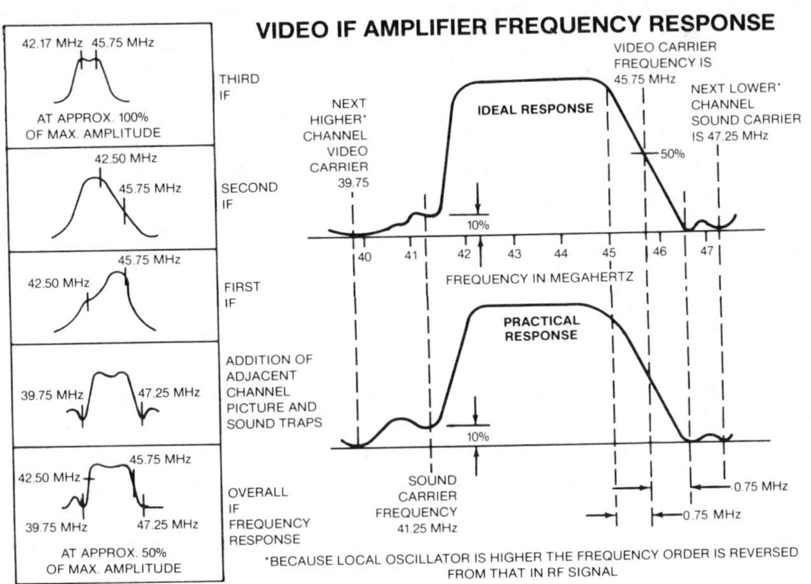

Obtaining an IF amplifier with such a frequency response is not as difficult as you might believe. The shape of the response of any individual stage can be controlled to some degree by adjusting the amount of coupling between primary and secondary windings of the interstage transformer. Damping resistors across inductors reduce the circuit Q and broaden the amplifier passband. The shape can be modified even more by using several IF stages with the interstages tuned to slightly different frequencies in the IF band. Such a tuning configuration is called *staggered tuning.*

IF Amplifier—Video Detector

The IF amplifier has three stages (Q101, Q102, and Q103). Stages Q101 and Q103 are common-emitter stages, while Q102 is in a common-collector configuration. Overcoupling of transformers, stagger tuning, and traps are used to shape the IF passband. As shown in the schematic diagram on the following page, the input from the tuner is applied to the base of the last IF stage (Q101). In addition to the input tuning (L101), several wave traps are present to reduce adjacent channel interference and to provide passband shaping. The series-tuned trap L107–C125 is tuned to the lower adjacent channel video (39.75 MHz) to suppress this signal. In like manner, L105, C104, and C105 are tuned to the higher adjacent channel to suppress the adjacent channel audio carrier. Series-tuned traps L108–C122 and L106–C123, provide for additional shaping of the IF passband. As shown, Q101 is neutralized by negative-feedback capacitor C129 to minimize tuning changes with varying AGC voltage. AGC is applied to the base of Q101 through R101 as shown. As the AGC voltage becomes increasingly positive (increasing signal level), Q101 draws more current, increasing the voltage drop across R105 (1.8 K). This lowers the collector voltage on Q101, reducing its gain proportionately as the AGC voltage increases. This AGC scheme is the same as that used for the RF amplifier in the UHF tuner, except that it is not delayed. You will study the AGC circuits later in this section. Q101 is coupled to Q102 via tuned interstage L114 in parallel with C110 and the circuit stray capacitance. Resistances R106 in parallel with R110 reduce the Q of this interstage, providing broadbanding. Q102 provides impedance matching between Q101 and Q103, having a high input impedance and a low output impedance (common collector). The Q103 output is a double-tuned circuit (L103–L104) that couples the output of Q103 to the detector CR101, as well as providing some passband shaping. L103 and L104 are capacitor-coupled to form an overcoupled transformer that provides for a wide passband with good selectivity. The output of the video detector diode CR101 is negative-going video. The IF signal is filtered out by the LC network shown, at the output of CR101. The video-detector filter network uses relatively short time constants, as you can see. This is necessary because the detector must pass video frequencies that extend up to 4 MHz and in fact must also pass the 4.5-MHz sound carrier that will be FM demodulated. The negative-going video output from the detector contains the AM video information. In addition to demodulating the AM video signal, the diode detector acts as a mixer that heterodynes the picture carrier against the sound carrier. This produces a difference frequency (4.5 MHz), which is the sound IF carrier. Note that this sound IF carrier is always at 4.5 MHz regardless of the local oscillator tuning. The 4.5-MHz sound carrier is amplitude modulated by the TV picture carrier. This AM is stripped off the FM sound carrier by limiting before FM demodulation. Sometimes, the limiting is not good enough, and you can hear a sound buzz that corresponds to the 60-Hz field rate in the sound channel.

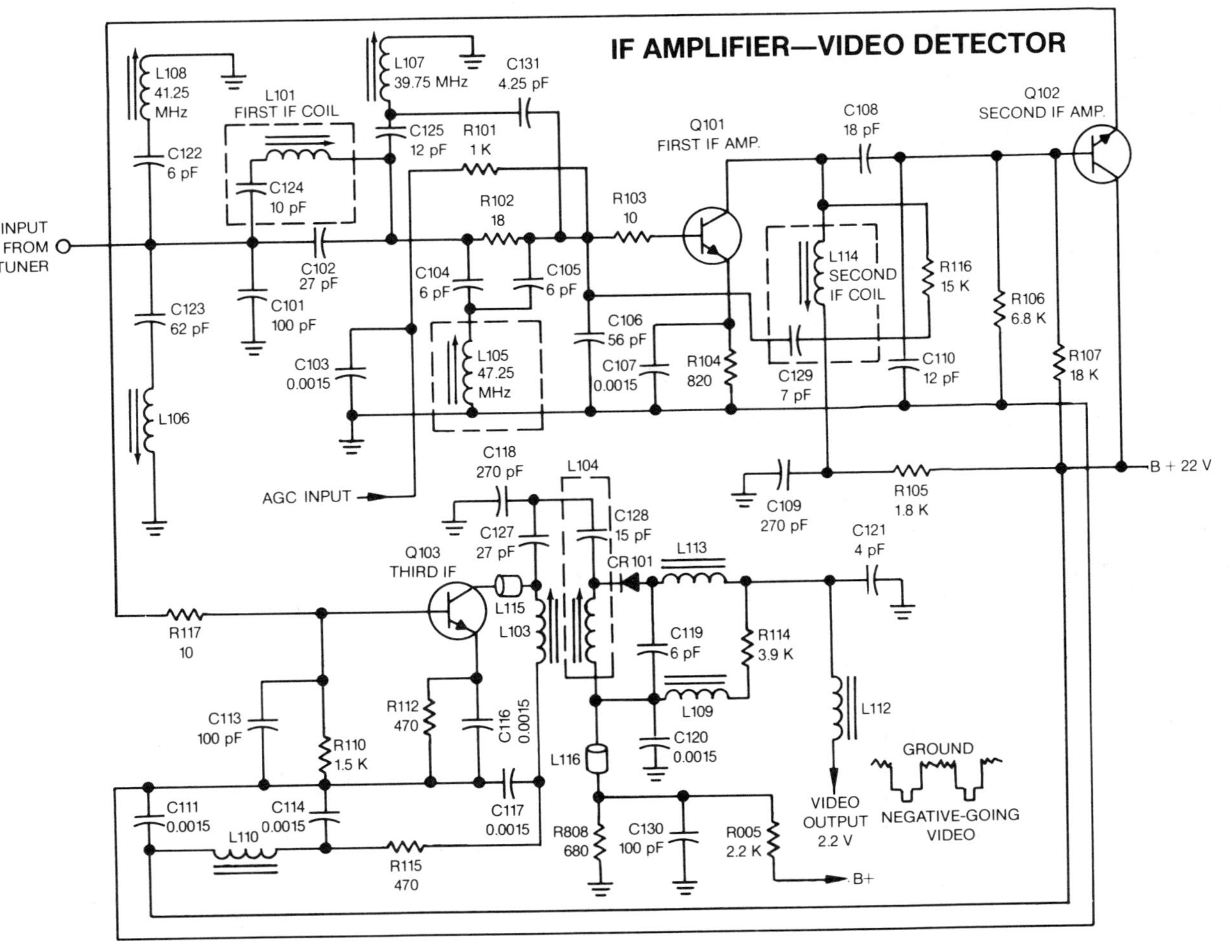
IF AMPLIFIER—VIDEO DETECTOR
L108 41.25 MHz
L101 FIRST IF COIL
L107 39.75 MHz
C131 4.25 pF
C125 12 pF
R101 1 K
C122 6 pF
C124 10 pF
INPUT FROM TUNER
C102 27 pF
C101 100 pF
C123 62 pF
L106
R102 18
C104 6 pF
C105 6 pF
L105 47.25 MHz
C103 0.0015
AGC INPUT
R103 10
Q101 FIRST IF AMP.
C106 56 pF
C107 0.0015
R104 820
C108 18 pF
L114 SECOND IF COIL
C129 7 pF
R116 15 K
C110 12 pF
R106 6.8 K
Q102 SECOND IF AMP.
R107 18 K
B + 22 V
C109 270 pF
R105 1.8 K
C118 270 pF
L104
C127 27 pF
C128 15 pF
Q103 THIRD IF
L115
L103
CR101
L113
C121 4 pF
R117 10
C119 6 pF
R114 3.9 K
L109
L112
C113 100 pF
R112 470
C116 0.0015
R110 1.5 K
C120 0.0015
L116
GROUND
C111 0.0015
C114 0.0015
L110
C117 0.0015
R808 680
C130 100 pF
R005 2.2 K
VIDEO OUTPUT 2.2 V
NEGATIVE-GOING VIDEO
R115 470
B+

Video Driver—Sync Separator

The detected negative-going video is applied to the video driver Q801. This signal contains the 4.5-MHz sound IF carrier as well as the video information. As shown in the diagram below, Q801 is an emitter follower. Thus the signal at the emitter of Q801 is also negative going. The video output from the emitter of Q801 is used to drive the video output (see page 4–136) and to drive the sound IF circuits. In addition, a sample of the negative-going video attenuated by a resistor network is applied to the AGC gate, which will be discussed later as part of the AGC system.

The sync-separator transistor is normally cut off but is driven into conduction by the positive-going video signal. The current flowing in the base/emitter circuit causes it to act like a diode that back-biases R/C so that only the very tips of the video (the sync pulses) drive the sync separator into conduction. The output consists of the stripped sync pulses. Note that the base bias automatically adjusts so that only sync pulses appear at the output, and these pulses are negative going.

THE SYNC SEPARATOR

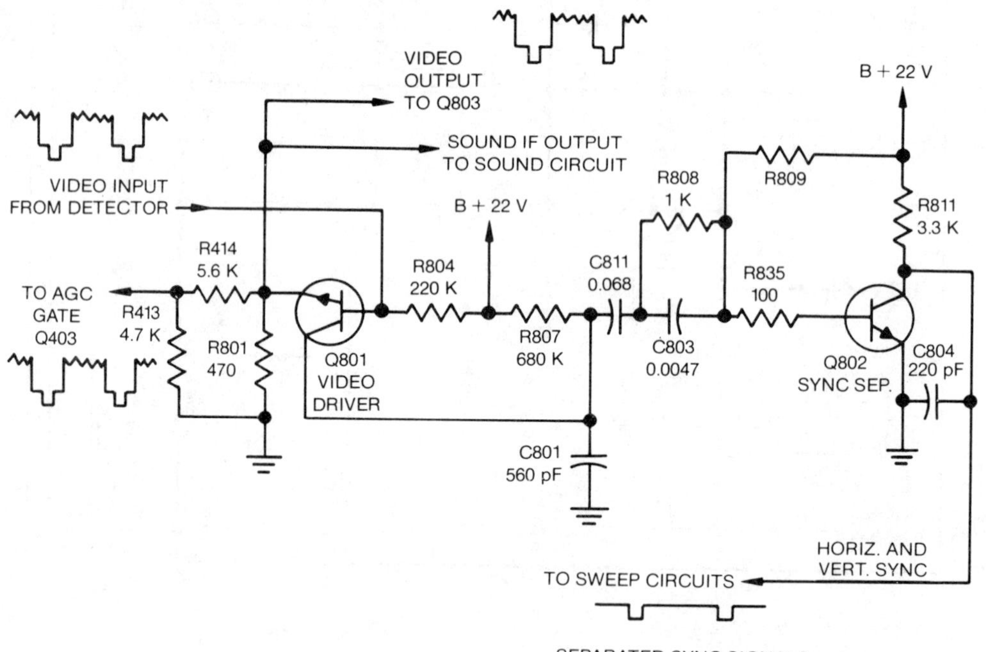

Video Driver—Sync Separator (continued)/Video Output—CRT Circuits

For the circuit shown on the previous page, a small resistance (R807) in the collector of video driver Q801 provides an inverted (positive-going) video signal that is capacitively coupled to the base of sync separator Q802. Thus, these positive pulses applied to the base of Q802 cause current to flow in the base-emitter junction of Q802. This back-biases the base of Q802 so that it is cut off except for the most positive part of the video waveform, the sync pulses. Thus, the collector output from Q802 consists of the sync pulses only, separated from the video signal, and these are used to provide synchronization of the horizontal and vertical sweeps.

Now let us move onto the study of video output/CRT circuits.

The video output/CRT circuits are shown on the next page. The negative-going video from the video driver (Q801) is applied to the base of the video output amplifier Q803 through contrast control R813. R813 is a voltage divider that adjusts the video level into the Q803 base. Since R813 controls the video level, it controls the peak-to-peak amplitude of the signal applied to the CRT cathode. Therefore, it controls the ratio of minimum to maximum picture contrast. Q803 is a common-emitter amplifier stage; therefore its output at the collector is positive going. Since these positive-going signals must reduce the beam current, the video is applied to the CRT cathode. The collector supply for Q803 is +180 volts, obtained from the high-voltage power supply to allow for generating the large video voltage (50–100 volts) necessary to fully modulate the CRT beam. The video signal is capacitance-coupled via C807 to the CRT cathode. The brightness control R829 adjusts bias on the CRT cathode to set the average intensity of the beam.

The CRT uses a fixed-focus voltage on the focus electrode provided by voltage divider R824–R832. The grid and second anode are operated at a fixed voltage provided from the 180-volt supply. The capacitor/resistance network momentarily holds the 180 volts of dc when the set is turned off to prevent a bright spot from appearing when the sweeps collapse, thus preventing the fluorescent screen from being burned from excessive beam current.

A parallel-tuned trap (at 4.5 MHz) in the emitter of Q803 provides a high impedance at the sound IF to reduce the gain of Q803 at 4.5 MHz to prevent sound signals from appearing at the video output. To ensure positive blanking during retrace, positive vertical- and horizontal-retrace blanking pulses from the sweep circuits are applied to the emitter of Q803. These form large positive-going blanking pulses at the collector of Q803, to ensure adequate blanking at all times.

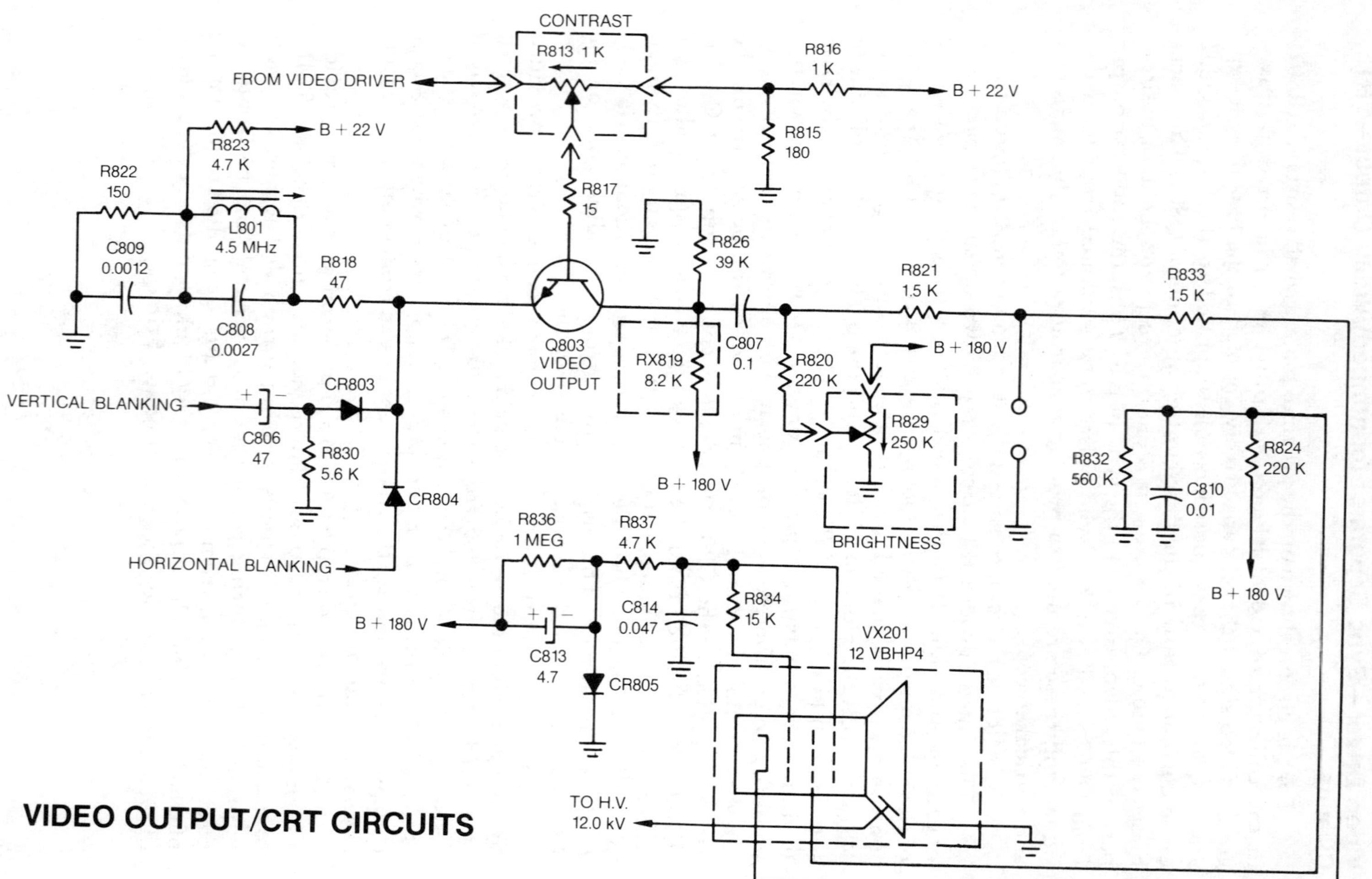

VIDEO OUTPUT/CRT CIRCUITS

AGC Circuits

TV AGC circuits are a little more complicated than a simple AGC for various reasons (see 4–39). The video signal during the blanking-pulse interval can be used for determining AGC. To do this, the video signal must be gated during blanking into a suitable peak detector that holds the peak blanking level so that a control voltage proportional to signal strength is generated for AGC—that is, proportional to the video amplitude during the blanking/sync-pulse interval (see figure).

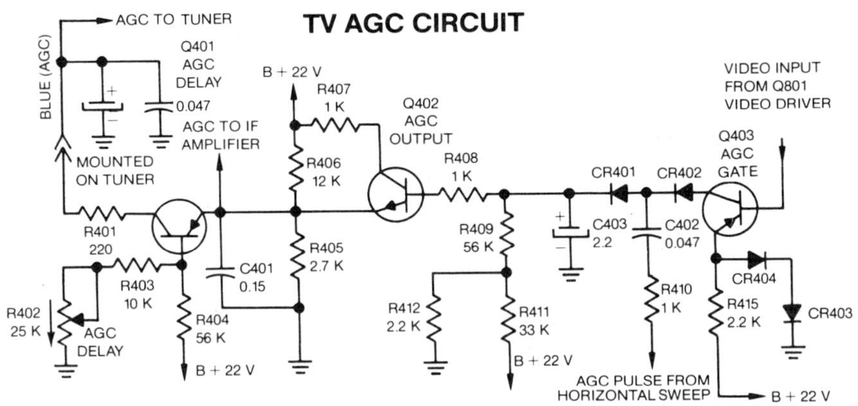

The AGC gate (Q403) allows the video from the video driver to be gated through during the retrace time of the horizontal sweep. As shown, Q403 has its emitter clamped to 1.3 volts dc by the forward-biased diodes CR403 and CR404. The base of Q403 is biased from the output network of Q801 (video driver) so that the negative video signals tend to drive Q403 into conduction. The collector voltage for Q403 is derived from the negative-going AGC pulse that occurs during retrace. Thus, the only time Q403 is active is during retrace. The video signal thus is sampled during the blanking/sync-pulse interval. Since the video input to Q403 is negative going, the video sample at its collector is positive going. This causes diode CR401 to conduct, storing a charge on capacitor C403 proportional to the peak video level. After many samples are stored, C403 represents the average signal level during the blanking/sync interval. The bias network in the base of the AGC ouput Q402 allows this charge to leak off slowly if no video signal is present. Q402 functions as an emitter follower with the dc voltage stored on C403 on its base. This dc voltage becomes increasingly positive as the signal level increases and is fed to the base of the first IF amplifier (Q101) to reduce its gain as the signal level tends to increase. The output of Q402 is also applied to the emitter of Q401 (common-base amplifier). Potentiometer R402 adjusts the base bias on Q401 so that it is normally cut off, and it is only when the emitter of Q401 becomes sufficiently positive that Q401 begins to conduct and provide AGC voltage to the RF amplifier. Thus, AGC action for the RF amplifier is delayed and is only applied when dealing with strong signals. In this way, full sensitivity is maintained.

Separating the Horizontal and Vertical Sync

As you learned, the sync-pulse train consists of a series of pulses at the horizontal-line rate of 15,750 Hz. These are narrow pulses about 4 μsec wide during all intervals except during the vertical-sync interval when these pulses are made to be almost the full horizontal period in width (about 60 μsec). However, the leading edges of the horizontal-sync pulses are always spaced precisely by the line interval, even though their width may vary. The variable width provides the basis on which the horizontal and vertical sync are separated from each other after being separated from the video signal. The separation of the sync pulses is accomplished by using RC circuits called *differentiators* and *integrator circuits*, as shown.

SEPARATION OF SYNC PULSES

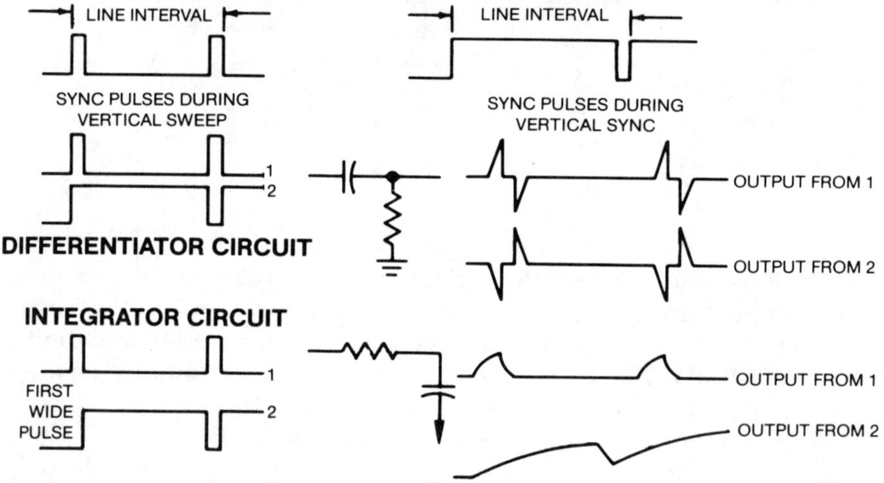

LINE INTERVAL
LINE INTERVAL
SYNC PULSES DURING VERTICAL SWEEP
SYNC PULSES DURING VERTICAL SYNC
1
2
OUTPUT FROM 1
DIFFERENTIATOR CIRCUIT
OUTPUT FROM 2
INTEGRATOR CIRCUIT
1
FIRST WIDE PULSE
2
OUTPUT FROM 1
OUTPUT FROM 2

A differentiator circuit has an RC time constant that is shorter than the pulse length. Thus, the output, as you know from your study of amplifiers, consists of the leading and trailing edges of the pulses, as shown. You will note that the leading edges of a uniform pulse train like the horizontal-sync pulses are the same regardless of the length of the intervening pulse. By passing the separated sync signal through a differentiator, a series of spikes representing the horizontal sync is obtained; these spikes are continuous, even during the vertical-sync interval. To recover the vertical sync, an integrator circuit is used. Here the time constant for the RC shown is long compared to the normal horizontal-sync pulse time. The capacitor of the RC cannot charge much during the horizontal-sync pulses, and discharges between them result in no buildup. On the other hand, during the vertical-sync pulse interval the horizontal pulses are very wide, and these do build up in the integrator to produce the vertical-sync pulse output, as shown.

In this way, the differentiator (short time constant) or high-pass filter separates the horizontal-sync pulses, while an integrator (long time constant) or low-pass filter separates the vertical-sync pulses.

Vertical-Sweep Circuits

The sync pulses separated by Q802 (sync separator) are negative going at the collector of Q802. These sync pulses are applied directly to the integrator circuit of the vertical sweep circuits as shown on the next page. The integrator consists of R625, C608, R624, and C607. The diode network CR601–CR604 acts as a clipper to prevent large noise pulses from triggering the vertical oscillator improperly. Diode CR601 is forward biased to a 0.5-volt level and acts as the return path for the emitter of Q601. Diode CR604 clips the negative peaks in excess of −0.5 volt.

The vertical oscillator is a collector-coupled multivibrator (Q601–Q602), shown in simplified schematic form in the diagram.

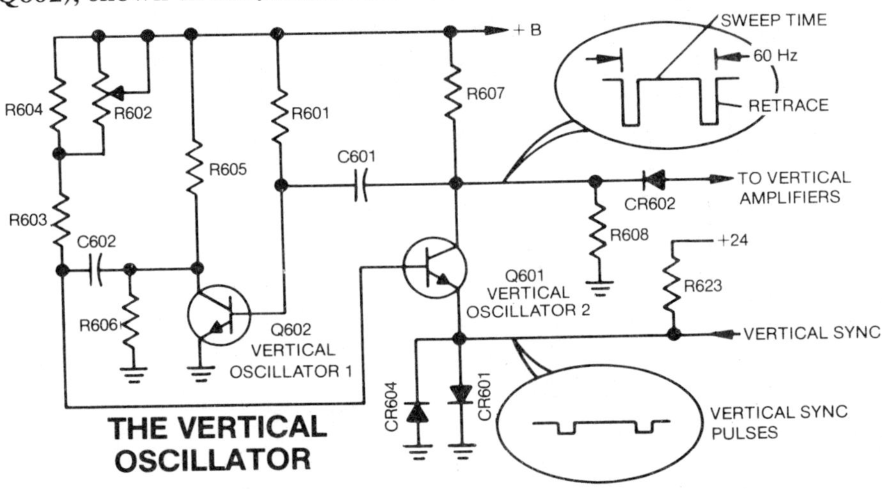

THE VERTICAL OSCILLATOR

Assume that Q602 has just become conducting via base current from R601. The negative-pulse output from the collector of Q602 is coupled to the base of Q601, cutting Q601 off. C602 recharges through network R604, R602, and R603, and when the voltage rises sufficiently, Q601 conducts and its collector voltage decreases. This negative-going waveform is coupled via C601 to the base of Q602, which is cut off until C601 recharges through R601. The time constant C601–R601 is very short compared to the time constant of R604, R602, R603, and C602; therefore, Q601 off time is much longer than its on time. This results in the waveform shown at the collector of Q601. While the on time of Q601 is determined by R601–C601, the off time is set by R602, R603, R604, and C602. Since the time constant of this network is variable (R602), the frequency of oscillation is adjustable by R602.

Negative-going sync pulses from the integrator cut off diode CR601. This causes Q601 to become nonconducting, thus forcing the cycle to be synchronized to the incoming vertical-sync pulses. As you can see, a high measure of mistriggering immunity is achieved because during the active time of the sweep (Q601 cut off), noise pulses cannot affect Q601 since it is already cut off. The vertical-sweep gate generated by the multivibrator is coupled through diode CR602 to the next stage of the vertical-sweep circuit, shown in the complete schematic.

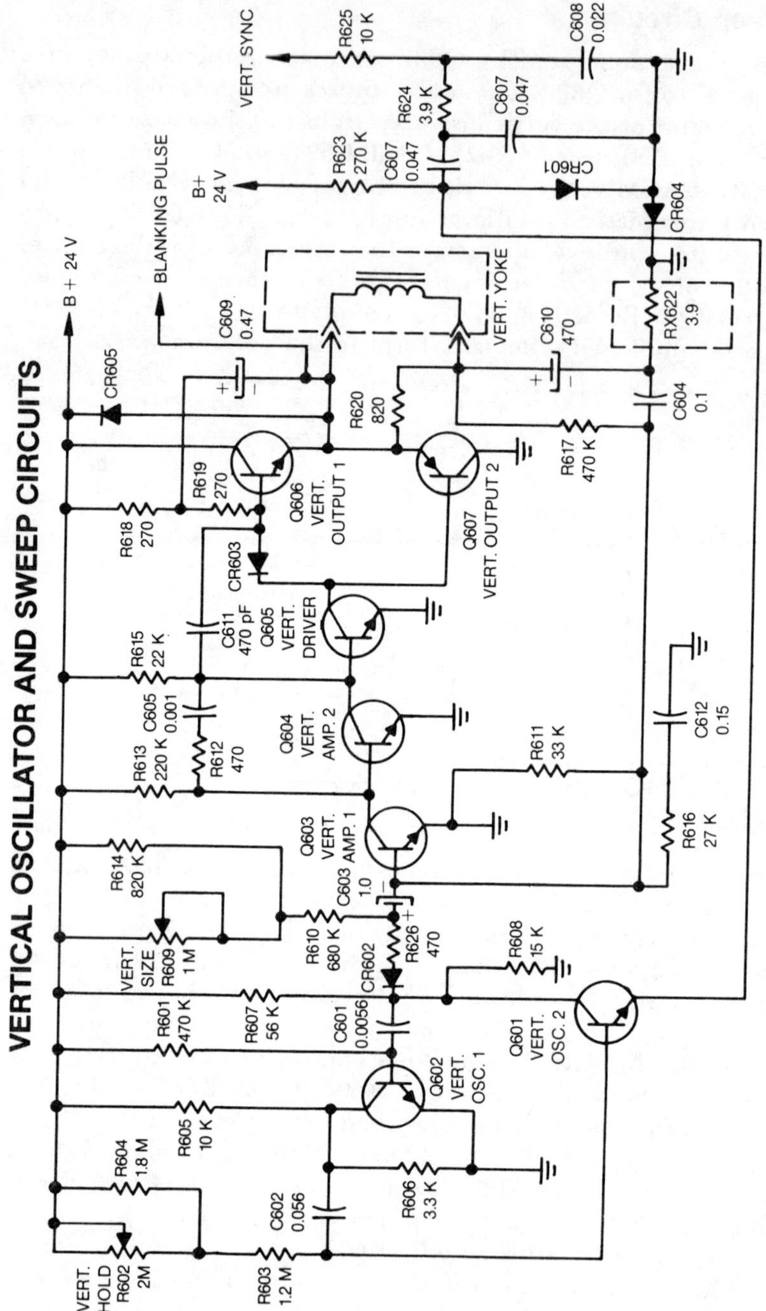
VERTICAL OSCILLATOR AND SWEEP CIRCUITS
B + 24 V
BLANKING PULSE
B+
24 V
VERT. SYNC
R625
10 K
C608
0.022
R624
3.9 K
C607
0.047
R623
270 K
C607
0.047
CR601
CR604
C609
0.47
VERT. YOKE
C610
470
RX622
3.9
CR605
R620
820
C604
0.1
R617
470 K
R618
270
R619
270
Q606
VERT.
OUTPUT 1
Q607
VERT. OUTPUT 2
CR603
C611
470 pF
Q605
VERT.
DRIVER
R615
22 K
C605
0.001
R613
220 K
R612
470
Q604
VERT.
AMP. 2
R611
33 K
C612
0.15
Q603
VERT.
AMP. 1
R616
27 K
R614
820 K
C603
1.0
VERT.
SIZE
R609
1 M
R610
680 K
R626
470
CR602
R608
15 K
R601
470 K
R607
56 K
C601
0.0056
Q601
VERT.
OSC. 2
Q602
VERT.
OSC. 1
R605
10 K
R604
1.8 M
C602
0.056
R606
3.3 K
VERT.
HOLD
R602
2M
R603
1.2 M

Vertical-Sweep Circuits (continued)

The complete schematic of the vertical-sweep circuits is shown on the previous page. The vertical-synchronized sweep gate from the collector of Q602 (the vertical oscillator) is applied to the Q603 base via CR602 and C603. When Q601 collector is positive, CR602 is nonconducting and C603 charges linearly through resistor network R609, R614, and R610, forming the vertical-sweep sawtooth. When Q601 becomes conducting, diode CR602 conducts, discharging capacitor C603 during the vertical retrace. Thus, a vertical sawtooth of voltage is formed. As you will remember, for a linear sweep, a sawtooth of current must flow through the deflection yoke. The current through the vertical deflection yoke is sampled by RX822. Thus, a voltage is generated across RX822 which represents the current flow through the vertical deflection yoke. This voltage is compared to the sawtooth generated at the input of Q603. The resultant is an error voltage applied to the vertical-amplifier circuits such that a sawtooth current is developed through the vertical-deflection yoke which will be linear and the same as the input sawtooth at Q603. The feedback amplifier consists of Q603, Q604, Q605, Q606, and Q607. Q603 amplifies and inverts the error voltage (sawtooth) waveform, which is then dc coupled into amplifiers Q604 and Q605. The negative-going waveform at the collector of Q605 is applied to the bases of the complementary-symmetry transistor pair Q606 and Q607. During the more positive portion of the waveform at the Q605 output, Q606 conducts with Q607 cut off, while during the more negative portion of the input waveform, Q607 conducts while Q606 is cut off. Diode CR603 smooths the switching action. As you can see, the output transistor pair is quite similar to the complementary-symmetry audio amplifiers that you studied earlier. The emitters of Q606 and Q607 are connected together and drive the vertical-deflection coils directly to provide the vertical sweep of the CRT trace.

As you can see, the basic vertical sync from the sync separator is used to determine the frequency of oscillation of the vertical-sweep oscillator, which in turn develops the vertical sweep. The vertical oscillator is free running so that the raster appears even where no signal is present. For synchronization to occur, it is necessary to have the free-running frequency of the oscillator slightly lower than the frequency of the sync signal. Otherwise the vertical oscillator will have started the next sweep before the sync pulse arrives.

Now that we have completed an in-depth study of vertical-sweep circuits we are ready to tackle our next subject—the study of horizontal-sweep circuits, the horizontal oscillator, and horizontal APC.

Horizontal-Sweep Circuits—Horizontal Oscillator—APC

The sync pulses from the sync separator are applied to the horizontal-sweep circuits via the differentiator, as shown in the partial schematic diagram on the next page. The diagram shows the horizontal oscillator and the horizontal APC (*automatic phase control*) that keeps the horizontal-oscillator frequency locked to the incoming sync pulses. The APC approach to horizontal-frequency control is used because it provides an extremely high degree of immunity to noise that can cause very annoying horizontal jitter. The noise immunity comes from the fact that the horizontal-oscillator frequency is controlled by averaging many samples so that oscillator frequency changes can only occur slowly. Because of limited pull-in range, the horizontal oscillator must be very stable in frequency, with only small corrections provided by the APC.

The horizontal oscillator Q502 is a conventional Hartley oscillator (base-emitter coupled) with its free-running frequency determined by inductor T501 and parallel capacitor C508. A portion of the oscillator voltage is coupled through C507 to the collector of Q501. Diode CR503 prevents the collector from swinging negative. Thus, the collector of Q501 consists of positive pulses. C507 can be considered to be in parallel with the horizontal-oscillator tuning capacitor C508. As the base voltage on Q501 varies, the amount of apparent capacitance from C507 varies, providing a fine adjustment of the frequency of the horizontal oscillator.

The input to Q501 base is obtained from the phase detector that compares the sync pulses to an APC pulse obtained from the horizontal flyback. The APC pulses are shaped into a sawtooth waveform by network RX506, C520, C502, and C503. The negative-going horizontal-sync pulses are applied to the diodes CR501 and CR502, causing them to conduct. These clamp a portion of the sawtooth generated from the APC pulse to ground. The point where this occurs depends on the phase relationship between the sawtooth and the sync pulse, producing a voltage at the input of R504 that depends on the point where the sawtooth is clamped to ground. Thus, varying voltage is filtered by R504, C506, R505, and C505 and applied to the base of Q501 to vary its collector impedance, which adjusts the frequency of Q502 to the value that leads to a voltage from the phase detector that locks the horizontal oscillator in phase with the sync pulse.

The schematic diagram of the horizontal-sweep circuits (less those portions previously discussed) is shown on page 4–145. As shown in this schematic, the output from the horizontal oscillator is applied to the base of horizontal driver Q502. This horizontal driver plus the horizontal output (Q504) will be discussed in detail on page 4–144. On the following page we have the schematic diagram for the phase detector and APC.

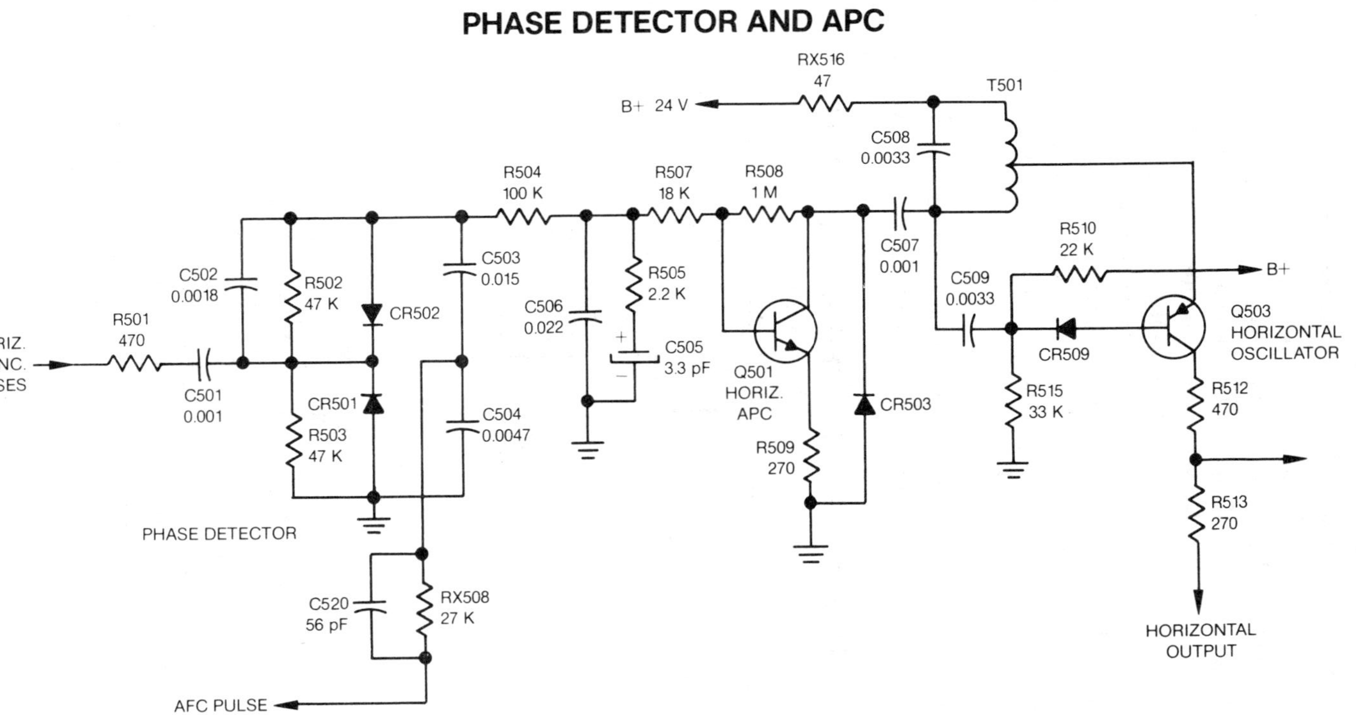
PHASE DETECTOR AND APC
RX516 47
B+ 24 V
T501
C508 0.0033
R504 100 K
R507 18 K
R508 1 M
C507 0.001
R510 22 K
B+
C502 0.0018
R502 47 K
CR502
C503 0.015
R505 2.2 K
C506 0.022
C505 3.3 pF
+
−
C509 0.0033
Q503 HORIZONTAL OSCILLATOR
R501 470
HORIZ. SYNC. PULSES
C501 0.001
CR501
R503 47 K
C504 0.0047
Q501 HORIZ. APC
CR503
CR509
R515 33 K
R512 470
R509 270
R513 270
PHASE DETECTOR
C520 56 pF
RX508 27 K
AFC PULSE
HORIZONTAL OUTPUT

Horizontal-Sweep Circuits—Horizontal Driver and Output

The voltage swing at the base of Q502 is sufficient to drive it between cutoff and saturation. Thus the collector voltage of Q502 consists of a square-wave signal at the horizontal frequency. This signal is applied to the base of the horizontal output Q504 via transformer T502. Q504 is normally cut off (base and emitter at ground potential). The signal from T502 drives Q504 into conduction (saturation), which causes transformer T503 to become saturated, causing increasing current through the transformer primary and the deflection yoke, deflecting the beam from the center of the screen to the right-hand edge. When Q504 is cut off, the magnetic field in the yoke collapses, causing the beam to retrace to the left-hand edge of the screen. The magnetic field in the yoke begins to build up as the transformer unsaturates, deflecting the beam toward the center of the screen, as shown by the waveforms below.

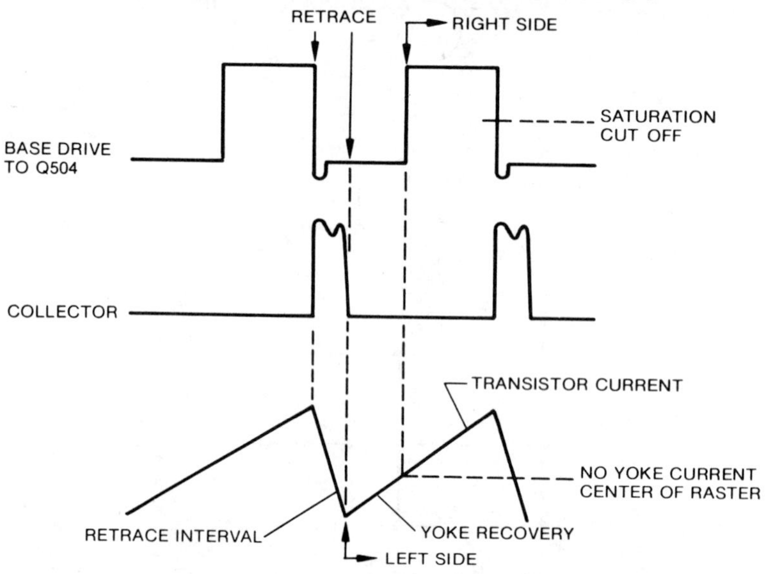

The inductive circuit formed by the deflection yoke and transformer T503 primary in parallel provides an inductive kick when the horizontal-output transistor Q504 is abruptly cut off, yielding the waveform at the Q504 collector shown above. When Q504 is conducting, current is drawn through the deflection yoke to provide the sweep. When Q504 cuts off, a 100-volt pulse appears at its collector. This is caused by the collapse of the magnetic field of the yoke, resulting in retrace of the horizontal sweep. This pulse is also used by T503 to provide high voltage, as described on the next page. A sample of the horizontal-retrace pulse is taken from the collector of Q504 via RX518 to the emitter of the video output transistor to provide horizontal blanking, as described earlier.

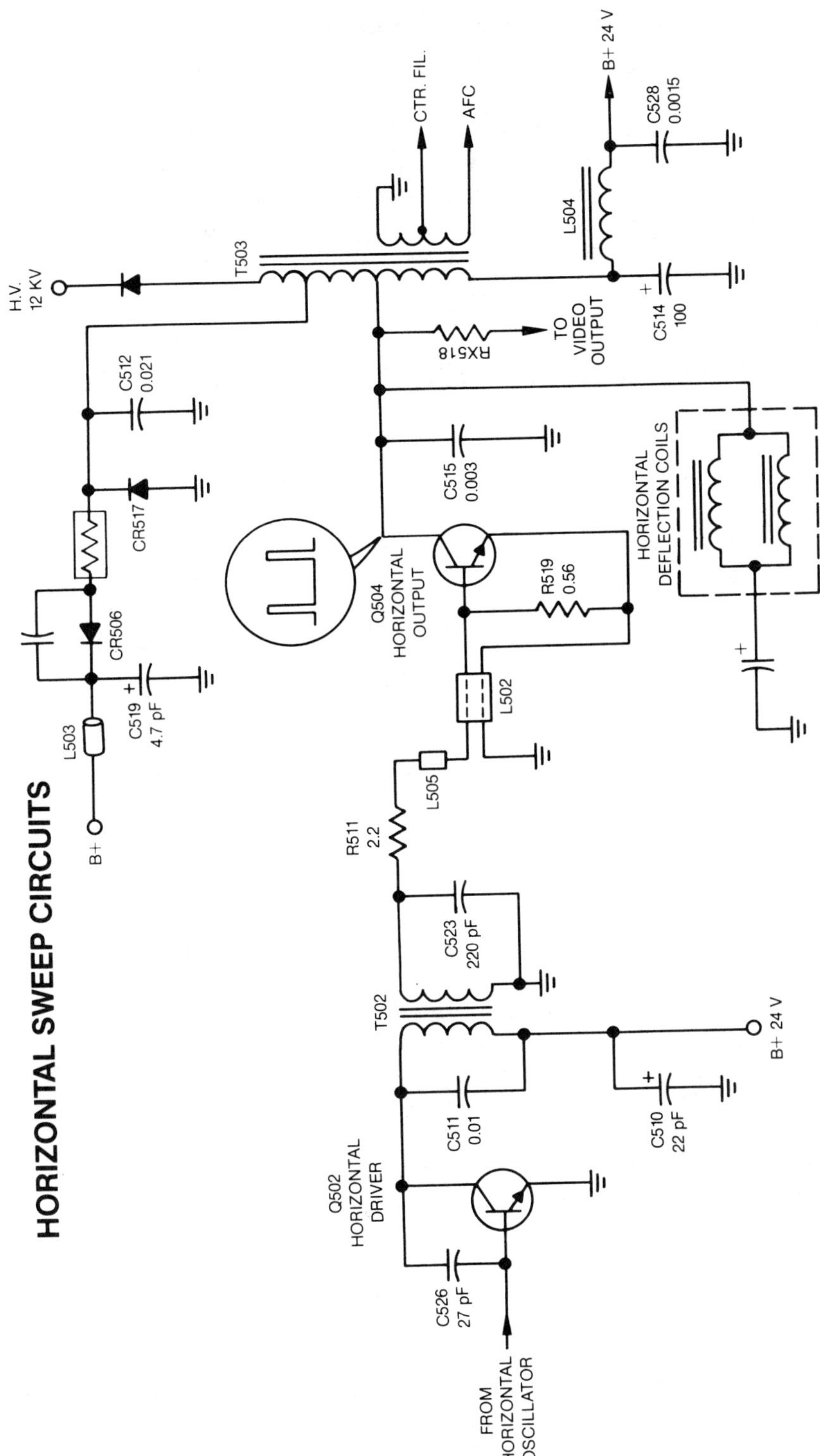
HORIZONTAL SWEEP CIRCUITS
B+
L503
C519
4.7 pF
CR506
CR517
C512
0.021
H.V.
12 KV
T503
CTR. FIL.
AFC
B+ 24 V
C528
0.0015
L504
C514
100
TO
VIDEO
OUTPUT
RX518
C515
0.003
HORIZONTAL
DEFLECTION COILS
Q504
HORIZONTAL
OUTPUT
R519
0.56
L502
L505
R511
2.2
C523
220 pF
T502
B+ 24 V
C511
0.01
C510
22 pF
Q502
HORIZONTAL
DRIVER
C526
27 pF
FROM
HORIZONTAL
OSCILLATOR

High- and Low-Voltage Power Supplies

The high voltage for the CRT second anode (12 kV) and +180 volts for operation of some of the video circuits and vertical-sweep circuits as well as CRT bias are obtained from T503. As shown in the diagram (see previous page for component value), the 100-volt retrace pulses at the collector of Q504, the horizontal output, are stepped up in the autotransformer primary of T503. The +180-volt output is obtained from an appropriate tap on T503 and rectified and filtered by CX506 and C519. Diode CRX517 clamps the bottom of the pulse waveform to ground so that the full pulse voltage available at the tap is obtained across C519.

HIGH- AND LOW-VOLTAGE POWER SUPPLIES

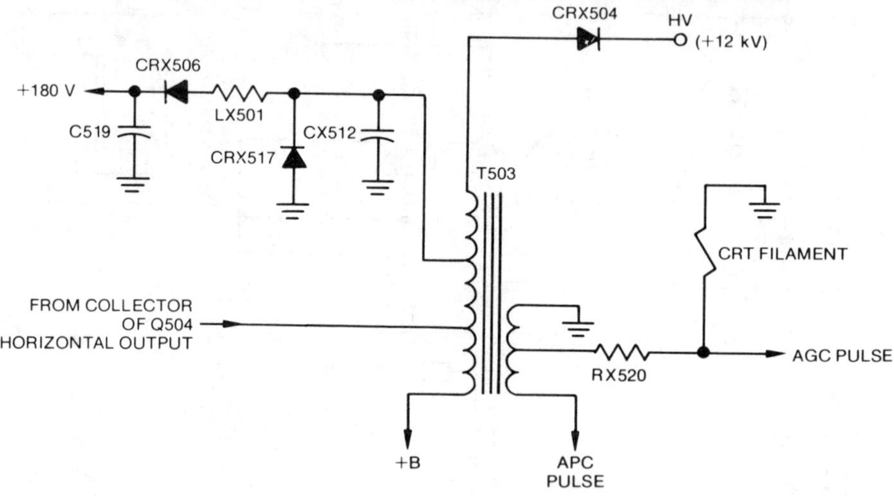

High-voltage dc (+12 kV) is obtained from the stepup of the collector pulse by autotransformer T503. This is rectified by diode CRX504. As described earlier, the high-voltage filter capacitor is formed by the final anode and an external conductive coating on the CRT envelope. Voltage for the CRT filament (a low-power filament) as well as the APC pulse and AGC gating pulse are obtained from a separate secondary on transformer T503. This method for obtaining filament voltage may seem unusual, but it is simple and convenient; in many TV sets, however, the filament voltage is taken from the low-voltage power supply transformer.

As shown by the schematic on the following page, the low-voltage power supply is of conventional design, providing an unregulated +24 volts of dc to the audio and sweep circuits. A value of +22 volts and a regulated +11 volts are provided for the video portion and the tuner, respectively.

For the circuit shown, a loss of sweep causes a loss of bias, but the bias voltage is held, as discussed earlier, long enough for the filament to cool so that no spot is formed. Otherwise, a disadvantage of the approach shown here would be that a failure of the sweep would also cause loss of CRT bias, leading to a bright spot that can burn the CRT screen.

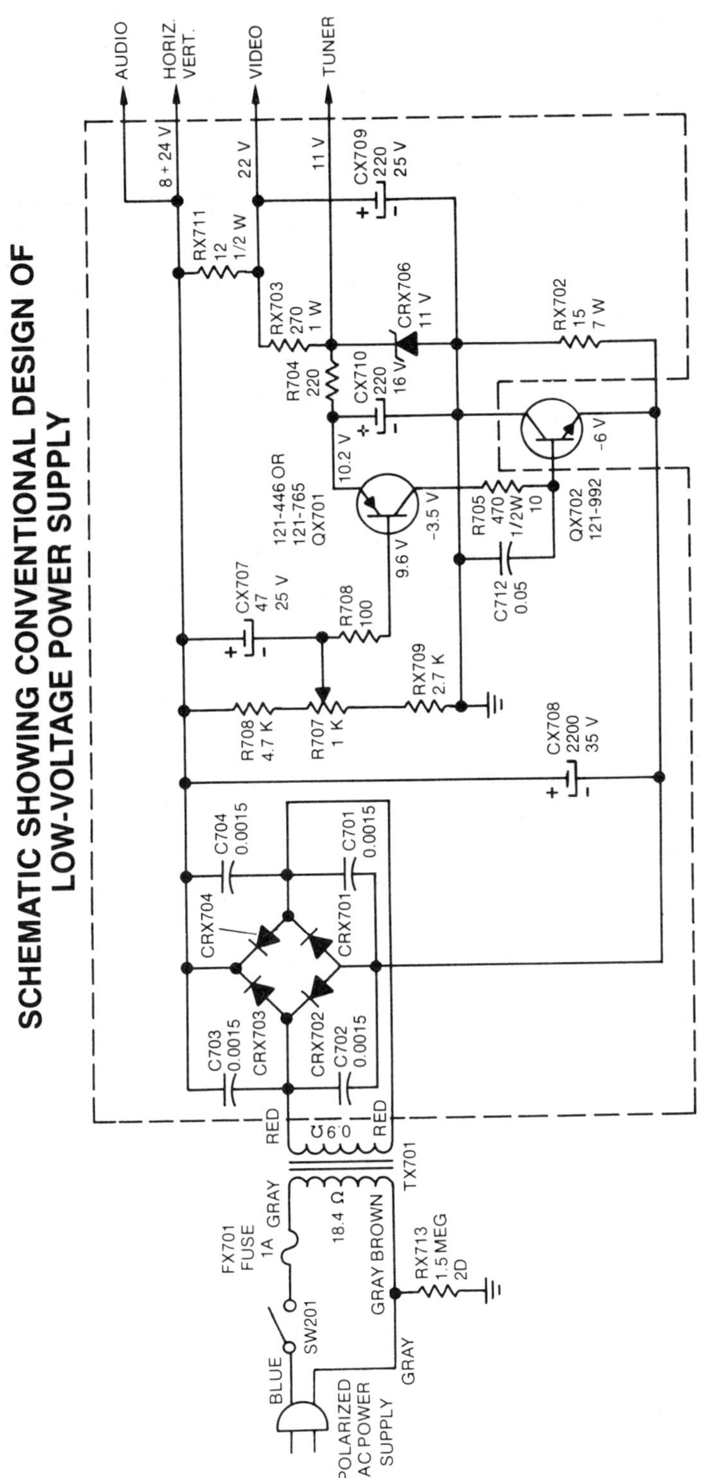

SCHEMATIC SHOWING CONVENTIONAL DESIGN OF LOW-VOLTAGE POWER SUPPLY

FM Sound Circuits

The sound circuits shown here and on the next page accept the 4.5-MHz IF carrier signal from the video driver Q801 and amplify and demodulate it in the same way as the FM radio receivers you learned about earlier in this volume. The resulting audio signal is amplified and used to drive a loudspeaker.

In the TV learning system, the sound IF demodulator system is contained in a single integrated circuit (IC1001), which also contains the ratio detector and first stages of audio amplification. The 4.5-MHz IF transformer (T1002) at the input to the IC and the quadrature coil (T1001) are external to the IC.

FM SOUND CIRCUIT

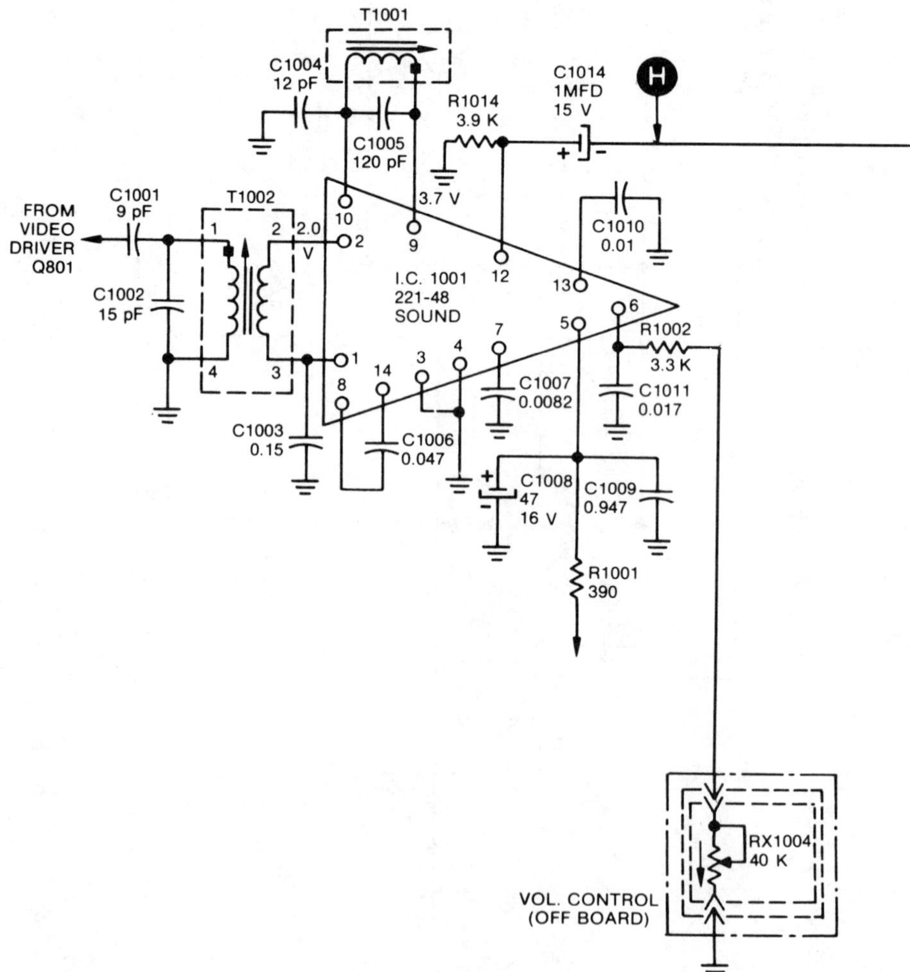

FM Sound Circuits (continued)

The input to IC1001 is the tuned transformer (T1002) coupled lightly via C1001 (8 pF) from the video driver Q801. The output of IC1001 is an audio signal that is fed to the audio amplifier. Volume is controlled by RX1004, which is connected to pin 6 of the IC. The audio output of IC1001 is fed to the complementary-symmetry amplifier (Q1001, Q1002, and Q1003) that you studied in Volume 2 of this series.

The audio output signal of the IC is developed across emitter resistor R1014, as shown. The audio power amplifier that receives the output signal from the IC consists of driver Q1001 and output transistors Q1002 and Q1003 connected in complementary symmetry. Q1003 is a pnp transistor, while Q1002 is an npn. They are connected as an emitter-follower combination that is biased to operate as a Class B amplifier. Use of this emitter-follower circuit permits connecting the output loudspeaker without the use of an output transformer.

FM SOUND CIRCUIT

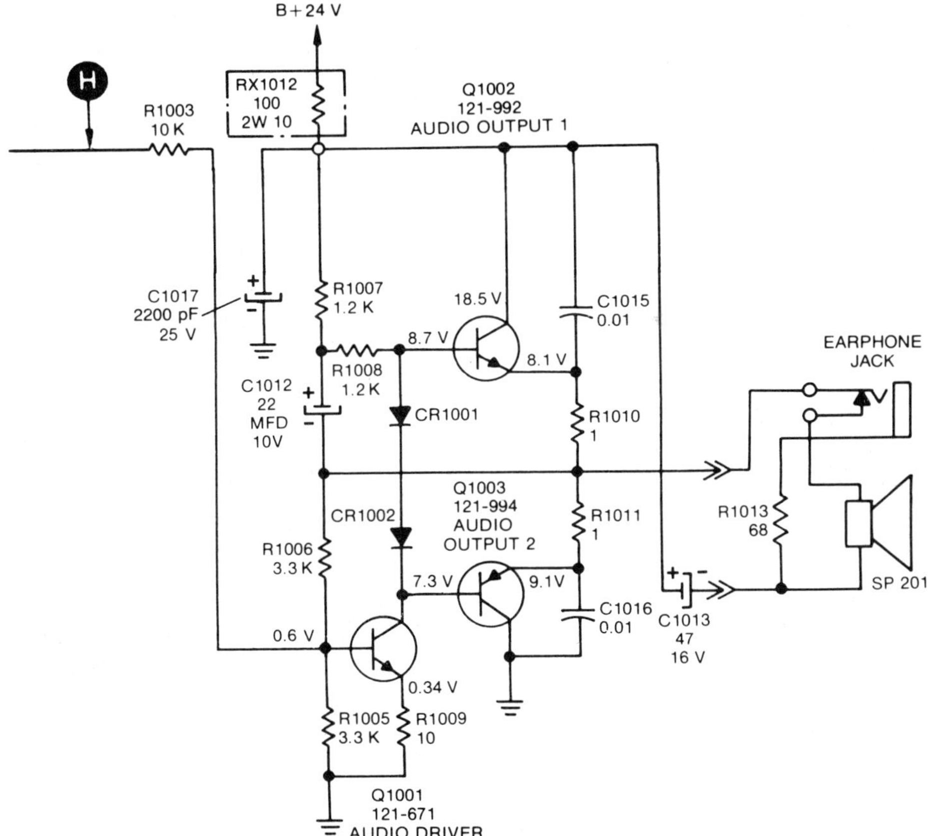

Review of Black-and-White TV Receivers

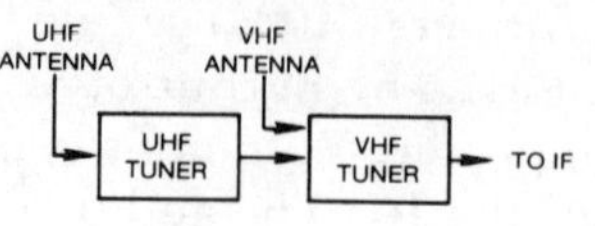

1. BOTH VHF AND UHF TUNERS are included in TV receivers. The output of the UHF tuner feeds through the VHF tuner.

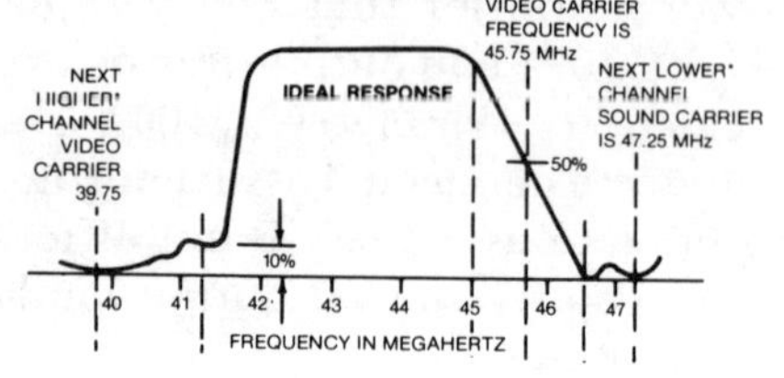

2. VIDEO IF AMPLIFIERS typically have three transistor stages and feed a diode video detector. They must have a response appropriate for proper video and sound reception.

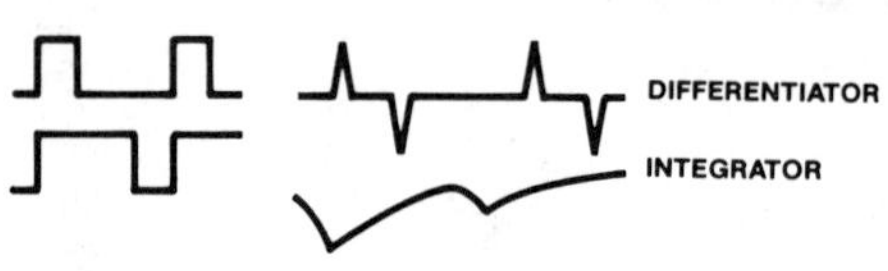

3. GATED AGC is provided by opening the AGC circuit only during horizontal sync pulses. Horizontal and vertical sync pulses are separated by a *differentiator* and an *integrator.*

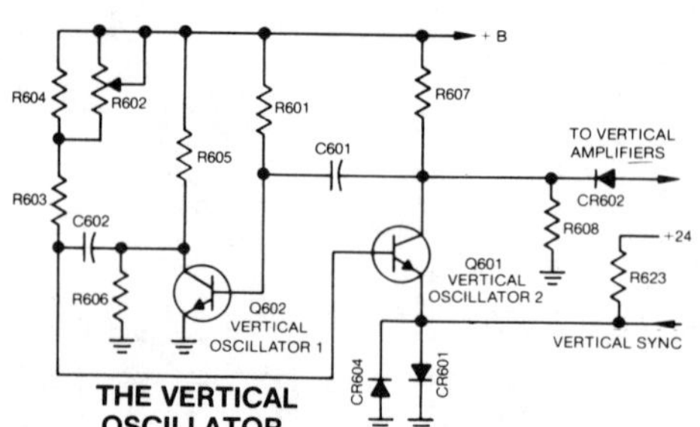

4. HORIZONTAL AND VERTICAL SWEEP CIRCUITS generate the voltages to deflect the CRT beam. Each has an oscillator and amplifiers. The horizontal sweep uses APC for noise immunity.

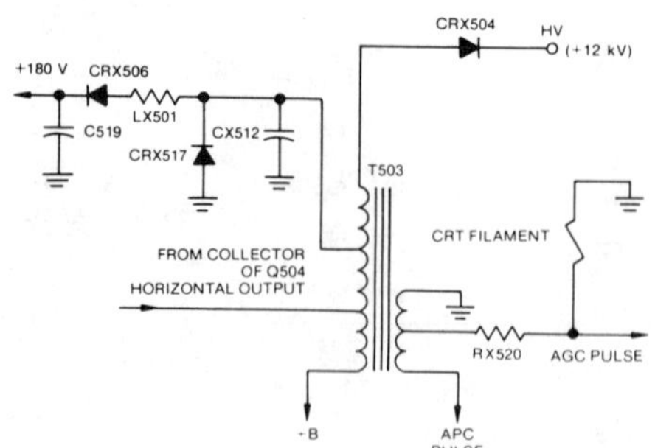

5. LOW VOLTAGE AND HIGH VOLTAGE POWER SUPPLIES are used. High voltage (and usually some low voltage) is obtained from the pulses in the *horizontal output transformer.*

Self-Test—Review Questions

1. Draw a complete block diagram of a black-and-white TV receiver.
2. Describe the function of each block from question 1.
3. Draw a typical TV channel showing location of video and sound carriers. What is the overall response shape of a TV video IF passband? How is this shaping accomplished? Why is it made this way?
4. Why is the sound carrier at 4.5 MHz regardless of where the receiver local oscillator is tuned? Sketch and describe the elements of a TV sound system.
5. What are some of the basic differences between a radio receiver RF/IF-detector system and a TV RF/IF-detector system? Explain. How are TV sets tuned?
6. How does a TV AGC differ from a radio receiver AGC? Why? Why is the tuner AGC delayed?
7. How are the sync signals recovered from the video signal? How are they separated?
8. Draw a block diagram of a horizontal-sweep circuit, including the horizontal oscillator and horizontal AFC. Describe how each element operates.
9. Draw a block diagram of the vertical-sweep circuits, including the sync circuits. Describe how each element operates.
10. How is high voltage for the CRT final anode obtained?

Learning Objectives—Next Section

Overview—Now that you have learned about black-and-white TV receivers, you are ready for the study of color TV receivers in the next section.

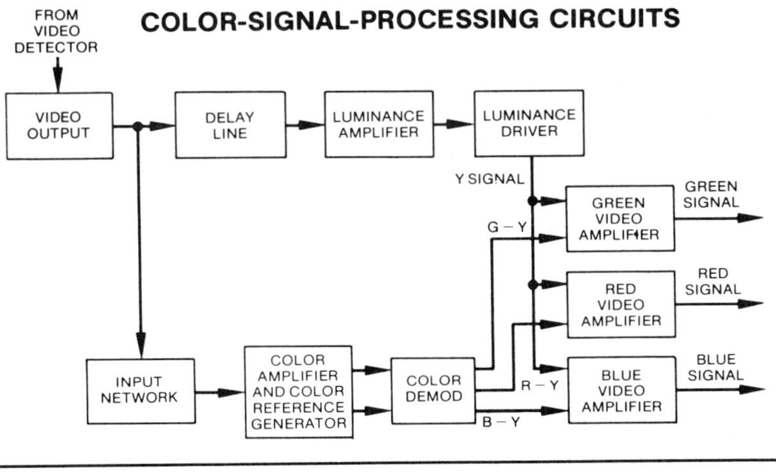

COLOR-SIGNAL-PROCESSING CIRCUITS

Introduction to Color Reception

You learned how a color TV signal was generated for transmission in your study of information transmission in Volume 3. Here, we will briefly review the form of the color signal. As you know, essentially any color can be made up from the three primary colors of red, green, and blue in the proper combinations. You also will recall that a given point in a TV scene can be described in terms of its *brightness* or *luminance* and its *hue* or *chrominance* (color). In black-and-white TV we used only the brightness or luminance signal. In color TV, a *second set* of signals is generated; these describe the *chrominance* and are transmitted on a *subcarrier*. The only real difference between a color signal and a black-and-white signal lies in the *additional information* contained in the *color subcarrier*. Thus, a color TV receiver is like a black-and-white receiver except for the special circuits to demodulate and regenerate the chrominance signal and the special CRT system to use this chrominance information to provide a color image.

COLOR TV RECEIVER

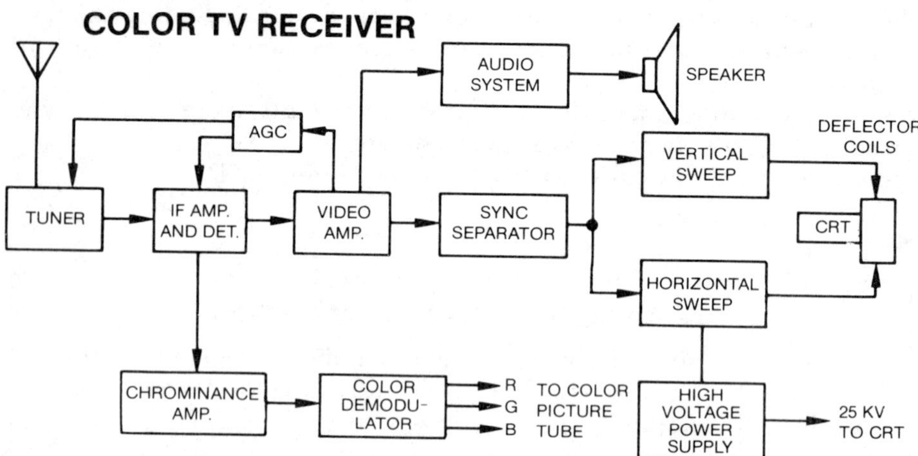

A block diagram of a color TV receiver is shown. As you can see, the RF, IF, audio, sync, and horizontal- and vertical-sweep circuits are essentially the same as for a black-and-white TV receiver. These circuits are similar to those you have already studied, except that more care is often used in the design of the IF circuits for color TV to ensure that all of the signals up to 4.2 MHz are received with good fidelity. In addition, the high voltage required for a color picture tube is generally higher (26–30 kV) and must be relatively stable, so that more careful design is required in this area as well.

The major difference lies in the video circuits, which include not only the necessary video amplifiers but also the circuits necessary to demodulate the chrominance signal and the additional circuits necessary to combine the luminance and chrominance information to recreate a color image on a color picture tube. Before you begin your study of the special circuits added to form a color TV receiver, the nature of the color TV transmission will be reviewed and the operation of the color picture tube or color CRT will be described.

The Color TV Signal

You will recall from your study of TV transmission that the color signals are generated at the studio by generating three signals using either three cameras with separate red-green-blue filters or by using a single tube with a special filter faceplate. These signals (E_R, E_G, and E_B for the red, green, and blue image signals, respectively) are then combined in a matrix, as shown, to form the luminance (or monochrome brightness) signal called E_Y. The E_R and E_B signals are also combined with the E_Y signals to form the E_I and E_Q video signals. E_I and E_Q are the chrominance signals, since they carry the color information. The E_I and E_Q signals are used to produce a suppressed-carrier double-sideband color-subcarrier signal centered at 3.579545 MHz, chosen at this frequency to ensure no interaction between the luminance and chrominance signals. The bandwidth of the color signals is carefully controlled, and the subcarrier is also carefully related to line and frame frequency.

COLOR TV SIGNAL

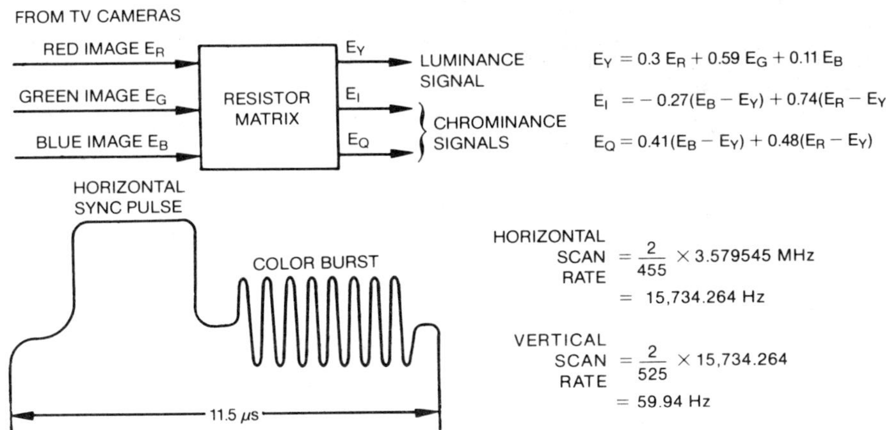

Since the chrominance information need not have very high resolution, because it does not carry the fine detail of the luminance information, the bandwidth of the E_I is limited to about 1.3 MHz and the E_Q to about 450 kHz. The band-limited E_I and E_Q signals are used to modulate the color subcarrier in quadrature, as will be described shortly.

To demodulate the suppressed-carrier chrominance signal, the carrier must be regenerated. Since phase is important, this carrier must be synchronized in phase with the subcarrier oscillator at the transmitter. To do this, a sample of the color subcarrier is transmitted on the back portion (back porch) of the horizontal-blanking pulses, as shown. This burst is recovered and is used to phase lock the receiver subcarrier oscillator. You will note that the horizontal- and vertical-scan rates are slightly different for color than for monochrome transmission. These differences are necessary to minimize any interference between the chrominance and luminance signals. These differences are so slight that the receiver can easily stay in sync when switching from monochrome to color or back. In fact, the old monochrome values are no longer used.

The Color TV Signal (continued)

The color information in E_I and E_Q must be combined so that they can be transmitted on the subcarriers, but in such a way that they can be recovered at the receiver. To understand how this is done, you will have to remember something about vectors and phase detectors that you also studied earlier. As shown in the diagram, the 3.58-MHz (3.58 MHz is often used as a shorthand way to express 3.579545 MHz) carrier is generated at the transmitter. It is used directly in the balanced modulator for the E_I signal, but it is phase shifted by 90 degrees for the balanced modulator for the E_Q signal. When the outputs from these balanced modulators are combined, the resultant is a single vector whose phase and amplitude depend on the relative magnitudes of E_I and E_Q. This single vector, containing both E_I and E_Q information, is the color signal that along with the E_Y and sync signals is used to modulate the video carrier. Obviously, if we can recover the E_Y, E_I, and E_Q signals at the receiver, we can put them into an appropriate matrix (add or subtract, as necessary) to reform E_R, E_G, E_B, and E_Y.

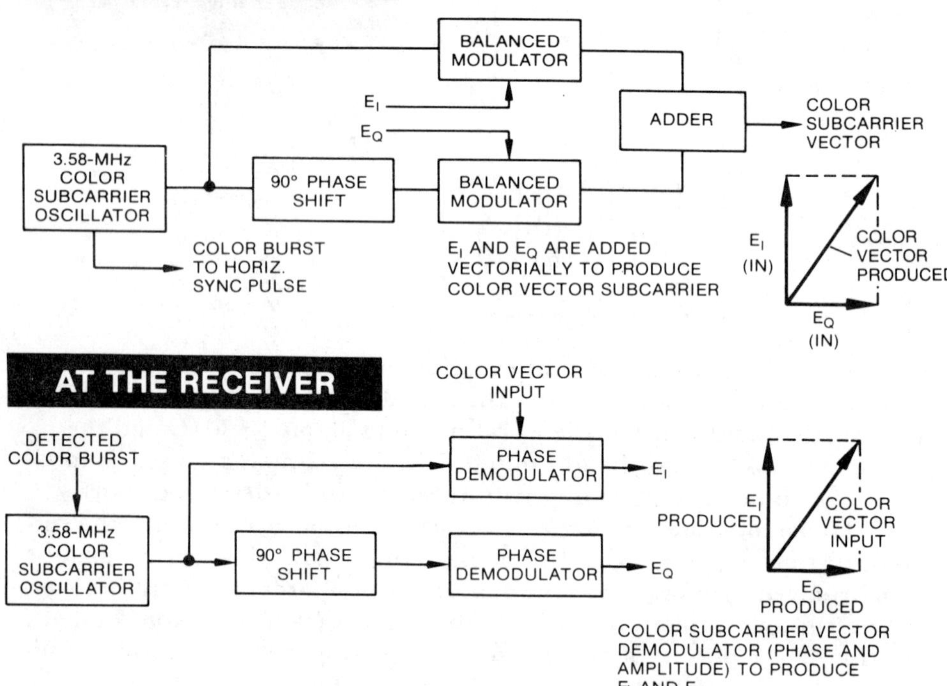

At the receiver, the color subcarrier is generated by locking a 3.58-MHz oscillator from the color bursts on the horizontal-sync pulses. This carrier signal is used directly in a demodulator (phase detector). The output from the phase detector is *only* the *in-phase* component of the color vector, which is E_I.

The Color TV Signal (continued)

As shown in the diagram on the previous page, an identical demodulator operated with a 3.58-MHz signal shifted 90 degrees produces only the E_Q (quadrature) component of the color vector. Thus, both E_I and E_Q are recovered from the color-subcarrier vector. The E_Y signal is recovered, as you have already learned, since it is the monochrome signal that is the same as a black-and-white video signal. Thus, the E_Y video, the E_I video, and the E_Q video are regenerated in the color TV receiver to produce the E_R, E_G, E_B signals necessary to operate the color TV picture tube.

As described on the previous page, the E_I, E_Q, and E_Y signals are recovered from the chrominance and luminance signals, respectively, to recover the red, blue, and green modulating signals. To do this it is now necessary to combine the E_I, E_Q, and E_Y signals in the right ratios and polarity. This is essentially what is done in a matrix similar to that used to generate the E_Y, E_I, and E_Q signals, which you learned about earlier. In most modern TV sets, this matrix operation is done in an IC containing the necessary resistors and inverters to reform the red, green, and blue signals (E_R, E_G, and E_B, respectively).

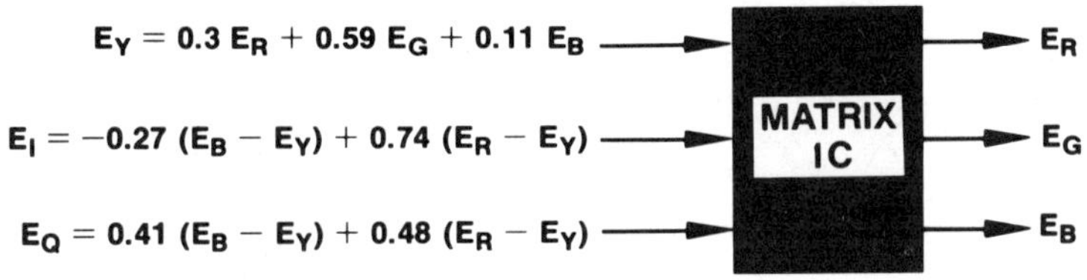

$$E_Y = 0.3\,E_R + 0.59\,E_G + 0.11\,E_B$$

$$E_I = -0.27\,(E_B - E_Y) + 0.74\,(E_R - E_Y)$$

$$E_Q = 0.41\,(E_B - E_Y) + 0.48\,(E_R - E_Y)$$

MATRIX IC → E_R, E_G, E_B

As you will see, the color picture tube has three electron guns that are designed to strike phosphors corresponding to a particular color—that is, a red gun that causes the screen to glow red, a green gun that causes the screen to glow green, and a blue gun that causes the screen to glow blue. You will note that at full intensity, the E_Y signal (monochrome) is equal to 1 ($0.3 + 0.59 + .11 = 1$). If this value is used to find E_I and E_Q, you will find that the sum of E_R, E_G, and E_B is equal to 1 also. Since we are dealing with difference signals, the difference $E_Y - (E_I + E_Q) = 0$, or all beams on full, giving a white picture if the CRT is properly adjusted. You will also find that each color can be recovered by substitution of the appropriate values in these equations. For example, if the picture is all blue, then $E_R = E_G = 0$. Substitution of these will show that only E_B remains. Similarly if $E_B = E_R = 0$, then only E_G remains; if $E_B = E_G = 0$, only E_R remains. You will see more clearly how these signals are used after you learn about the color picture tube.

It is important to realize that the red, blue, and green images are processed at the transmitter to form an E_Y, E_I, and E_Q signal, where E_Y is the monochrome signal and E_I and E_Q contain the information necessary, with E_Y, to form the red, blue, and green video signals. The recovery of E_Y, E_I, and E_Q at the receiver then likewise allows the recovery of E_R, E_B, and E_G at the receiver.

The Color Picture Tube

As you know, most colors can be generated by the appropriate combination of red, green, and blue (the three primary colors). If you look through a magnifying glass at a color TV set in operation, you will see that the color image is actually formed by clusters of tiny dots or strips of phosphor that glow red, green, and blue. These phosphor dots or strips are so small that they cannot be seen individually without a magnifier. A group of three spots (red, green, and blue) is called a *triad* and is somewhat smaller than an individual resolution element in the picture. If you watch these triads as the scene color changes, you will see that the relative brightness of the elements of the triad varies as the color changes from scene to scene. Since the phosphor dots or strips are so small, we see them collectively as a single spot whose color is determined by the *relative brightness* of the *individual elements*. The separate illumination of each of these spots is accomplished in a color TV picture tube using the three color signals E_R, E_G, and E_B described earlier.

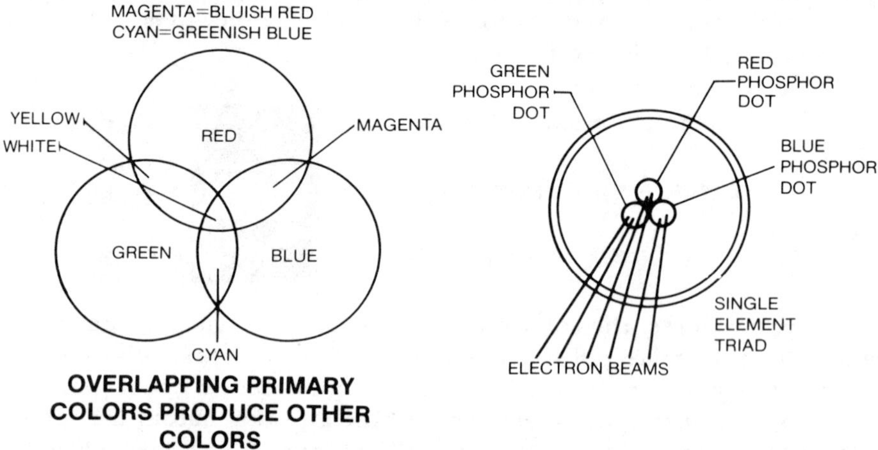

OVERLAPPING PRIMARY COLORS PRODUCE OTHER COLORS

Inside a typical color TV picture tube there are three electron guns, each producing a separate beam. These beams scan together, just as a single beam does in a monochrome picture tube. We can adjust the beam from each gun so that it strikes only a single kind of color phosphor dot—that is, the blue beam strikes only blue phosphor dots, the green beam strikes only green phosphor dots, and the red beam strikes only red phosphor dots. Then, if we adjust the *intensity* of each beam, we can produce in each triad *any color* we choose from white to any color to black (no beams on). We could receive a monochrome luminance signal and apply it to all three guns simultaneously after adjusting the ratio of the beam intensities to produce white light and have a satisfactory black-and-white image. This is what happens when we receive a monochrome picture on a color receiver. If we also individually modulate each beam at each point in accordance with the color information received, we would produce a color image, and this is exactly how color image is formed by driving each beam with its corresponding color video signal.

The Color Picture Tube (continued)

The basic problem in a color picture tube is making sure that the right beam strikes only the correct phosphor dots. An element known as the *shadow mask* is located behind the phosphor dot screen. Its purpose is to assure that the beam from the red gun passes through the shadow mask at an angle that will allow it to strike only the red-light-emitting phosphor. Similarly, the blue- and green-emitting phosphors can be struck only by the beams from the corresponding blue and green guns. This is accomplished basically by setting the guns and electron beams so that they converge through the holes in the shadow mask at an appropriate angle to give each beam the geometry necessary to permit it to strike only the appropriate color phosphor dots. The very fine registration necessary to achieve the purity of color by having each beam strike only its specific phosphor dots requires very careful control during the manufacture of the tube. Furthermore, most color TV sets use carefully regulated high voltage to maintain the registration and focus. Because some of the beam energy is lost by the shielding from the shadow mask, color CRTs usually use much higher final anode voltages to obtain a bright picture. In addition, since the beams cannot each occupy the same position in the deflection field, a set of adjustments to align each beam properly during deflection is necessary to maintain convergence of the beams to the proper color dots. These controls are called *convergence controls* and require resetting whenever the picture tube or its geometry is disturbed. The convergence controls in a conventional shadow mask color CRT are complex and difficult to adjust properly. Therefore, several of the newer color picture tube systems use somewhat different approaches to solve some of these problems. The phosphor dot screen is called a *P22 phosphor.* All modern color CRTs are rectangular with a 4/3 aspect ratio, as for black-and-white CRTs. Electrostatic focus is universally used.

OPERATION OF SHADOW MASK

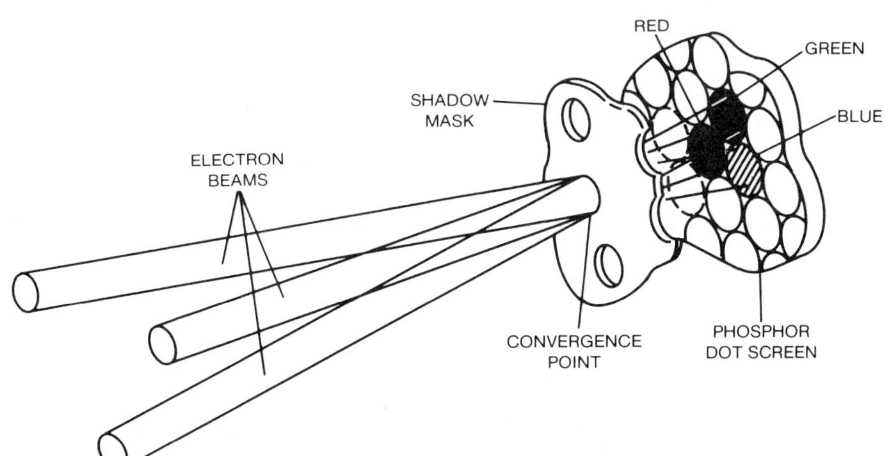

The Color Picture Tube (continued)

Several innovations have been used in color CRTs to reduce the critical aspect of the convergence controls and to simplify adjustment. One of these is the in-line gun structure developed by the Zenith Radio Corporation called the *Chromocolor*® picture tube. In this tube, the three guns are arrayed in a horizontal line. The phosphor is arranged in vertical strips and the shadow mask is also a vertical slit, so that the beams diverge through the slit to strike the color strips, as shown.

CONSTRUCTION OF COLOR PICTURE TUBE

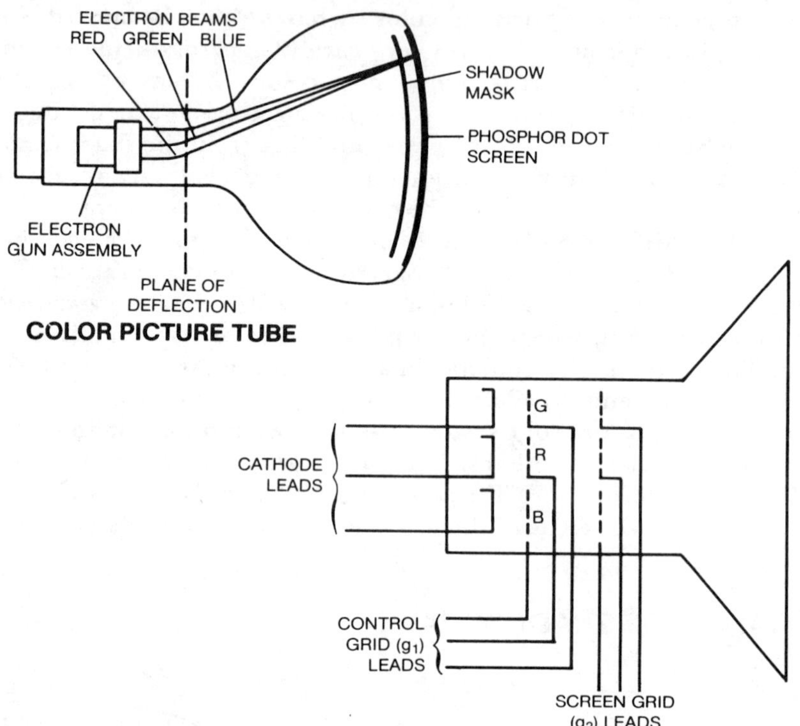

COLOR PICTURE TUBE

COLOR PICTURE TUBE SCHEMATIC

Since the convergence problems are limited to the horizontal direction only, a great simplification in adjustment is possible. Furthermore, the shadow-mask blocking area is reduced so that a brighter picture is possible for the same beam current.

The Color Picture Tube (continued)/Color Circuit Functions

Another innovation is the single-gun color picture tube. This tube uses a vertical strip phosphor system, just like the Chromocolor® tube, but the single beam is allowed to scan across all strips. Beam switching, as in the color single-gun camera tube, is used to separate the illumination of each vertical strip for each color.

In all color picture tubes, particularly in the larger sizes, the high voltage is sufficient to produce *soft x-rays*. While these are generally confined within the TV set to limits set by U.S. federal regulation, it is still a good idea to keep a *safe distance* (greater than 6 ft) when viewing a color TV receiver—particularly if the CRT is over 17 inches in size.

As you learned earlier, when the color TV signal is received, the processing is the same as for the black-and-white system, except for the *color-signal-processing circuits* not included in the black-and-white TV. The remainder of this section is devoted to describing those circuits.

The color-processing circuits in most modern TV receivers use ICs for many of the functions, so that the circuit functions are generally not obvious from the schematic diagrams. Therefore, in this section, while we will use a learning system approach as before, using the Zenith Radio Corporation model 19HC55, we will use a mixture of block diagrams and schematic diagrams to discuss the circuits unique to a color TV receiver.

COLOR CIRCUIT FUNCTIONS

1. EXTRACT COLOR INFORMATION FROM VIDEO
2. ADJUST INTENSITY OF BEAMS IN CRT
3. INDEPENDENTLY DRIVE THE R, G, B BEAMS
4. SUPPRESS COLOR INFORMATION DURING BLACK AND WHITE RECEPTION
5. PROPERLY CONVERGE CRT BEAMS

The color-processing circuitry of a TV receiver has *five functions*. First the *color information* must be *extracted* from the received signal, and the red, green, and blue *signals recovered*. Second, the *correct intensity* of each of the three electron beams must be *adjusted* so that the final color produced by the phosphor dots will *accurately duplicate* the colors in the original scene from the TV studio. Third, the red, green, and blue signals must be used to *drive* the three electron beams *independently*. Fourth, the colors must be *turned off* when only a black-and-white program is being received. Fifth, the three electron beams must always be made to *strike* the *correct dots* of the triad on the screen.

The pages that follow will describe how each of these objectives is achieved.

Color-Signal-Processing Circuits

The receiver color-signal-processing circuits extract the red (R), green (G), and blue (B) signals from the received signal. These are used to independently drive the three electron guns in the CRT. The block diagram shown here summarizes how this is accomplished in a typical color TV receiver.

The output of the video amplifier is the demodulated video signal, which is essentially a duplicate of the transmitted video. This signal contains the combined luminance (Y) and chrominance information, together with the blanking and sync signals and the color-sync burst.

COLOR-SIGNAL-PROCESSING CIRCUITS

FROM VIDEO DETECTOR
VIDEO OUTPUT
DELAY LINE
LUMINANCE AMPLIFIER
LUMINANCE DRIVER
Y SIGNAL
G − Y
GREEN VIDEO AMPLIFIER
GREEN SIGNAL
RED VIDEO AMPLIFIER
RED SIGNAL
INPUT NETWORK
COLOR AMPLIFIER AND COLOR REFERENCE GENERATOR
COLOR DEMOD
R − Y
B − Y
BLUE VIDEO AMPLIFIER
BLUE SIGNAL

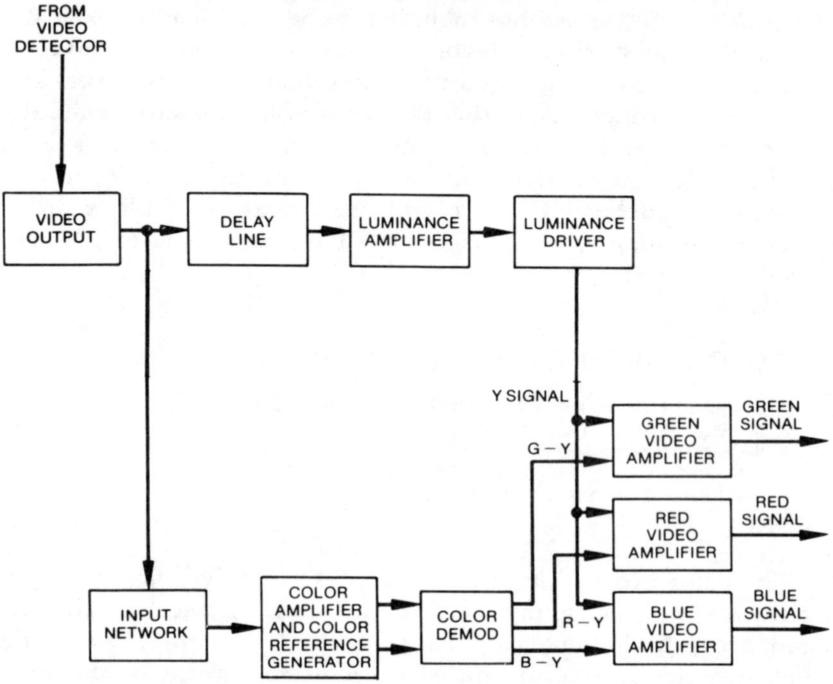

A 3.58-MHz wave trap at the output of the video amplifier is usually used to prevent the color subcarrier from entering the signal path to the luminance amplifier. A delay line at the input of the luminance amplifier delays the luminance signal in time so that it arrives at the output amplifiers *coincident* with the arrival of the corresponding chrominance signals, which are slightly delayed because of their narrower bandwidth and additional processing. After amplification in video amplifiers, the luminance signal is fed to the red-, green-, and blue-output amplifiers to be combined with R–Y, G–Y, and B–Y signals from the color or chroma sections.

Color-Signal-Processing Circuits (continued)/Video Output to CRT

The path of the chrominance signal is shown in the lower portion of the block diagram. The input signal to this path is the demodulated video signal, and it is first necessary to suppress the low-frequency components of the luminance signal. This is done by passing the video signal through a bandpass transformer that passes only signals in the 2.5- to 4.5-MHz range. The 3.58-MHz reference signal is generated by locking a local reference to the 3.58-MHz burst in the received signal. After amplification by the color amplifier, the signal enters the color demodulator where the R–Y, G–Y, and B–Y signals are recovered. The demodulator output is sent to individual color-output amplifiers, where they are each combined with the luminance signal for delivery to the three guns of the color CRT. As you can see, the Y signal is added to the R–Y, G–Y, and B–Y signals from the color section, yielding an R,G,B signal for modulating the color CRT. For black-and-white reception, the R–Y, G–Y, and B–Y signals are absent, leaving only the luminance signal Y to drive the three CRT grids.

In the discussion to follow here and on subsequent pages, we will start with the video ouput/CRT driver and work back to the video detector output at the IF. In this way, you will be able to see what signals are required at each step and how they are derived. You know that three signals (E_R, E_G, and E_B) are needed to drive the CRT guns. You know that the signals applied to these guns are simply proportions of the luminance (Y) signal for black-and-white reception. The schematic on the following page shows the video output circuits for the Zenith model 19HC55. The basic inputs, as shown, are the Y signal and the R–Y, G–Y, and B–Y signals.

As shown, the Y (luminance) signal is applied to the base of Q1201, the luminance driver. A resistance-divider network contains the precision resistors of the correct value to establish the magnitude of the Y signal into Q1201. All components that do not have identifying symbol numbers are in a separate IC. The output from the emitter of Q1201 is coupled to the emitter of the color-output amplifiers Q1205, Q1206, and Q1207 through resistor networks in U1202. Thus, Q1205, Q1206, and Q1207 act as common-base amplifiers for the Y signal. The outputs of these amplifiers are coupled to the cathodes of the CRT. The Y signal input is positive going, and since common-base amplifiers are noninverting, positive video is applied to the CRT cathodes. The R–Y, G–Y, and B–Y signals obtained from the color demodulators, to be described subsequently, are applied to the bases of the output amplifiers via resistive networks and emitter followers Q1202, Q1203, and Q1204. These negative-going signals at the bases of the output amplifiers are then added in the output of the output amplifiers. Thus, $(B - Y) + Y$ yields B, etc. In this way, the three color signals for the CRT are derived from the color video and the Y (luminance signal).

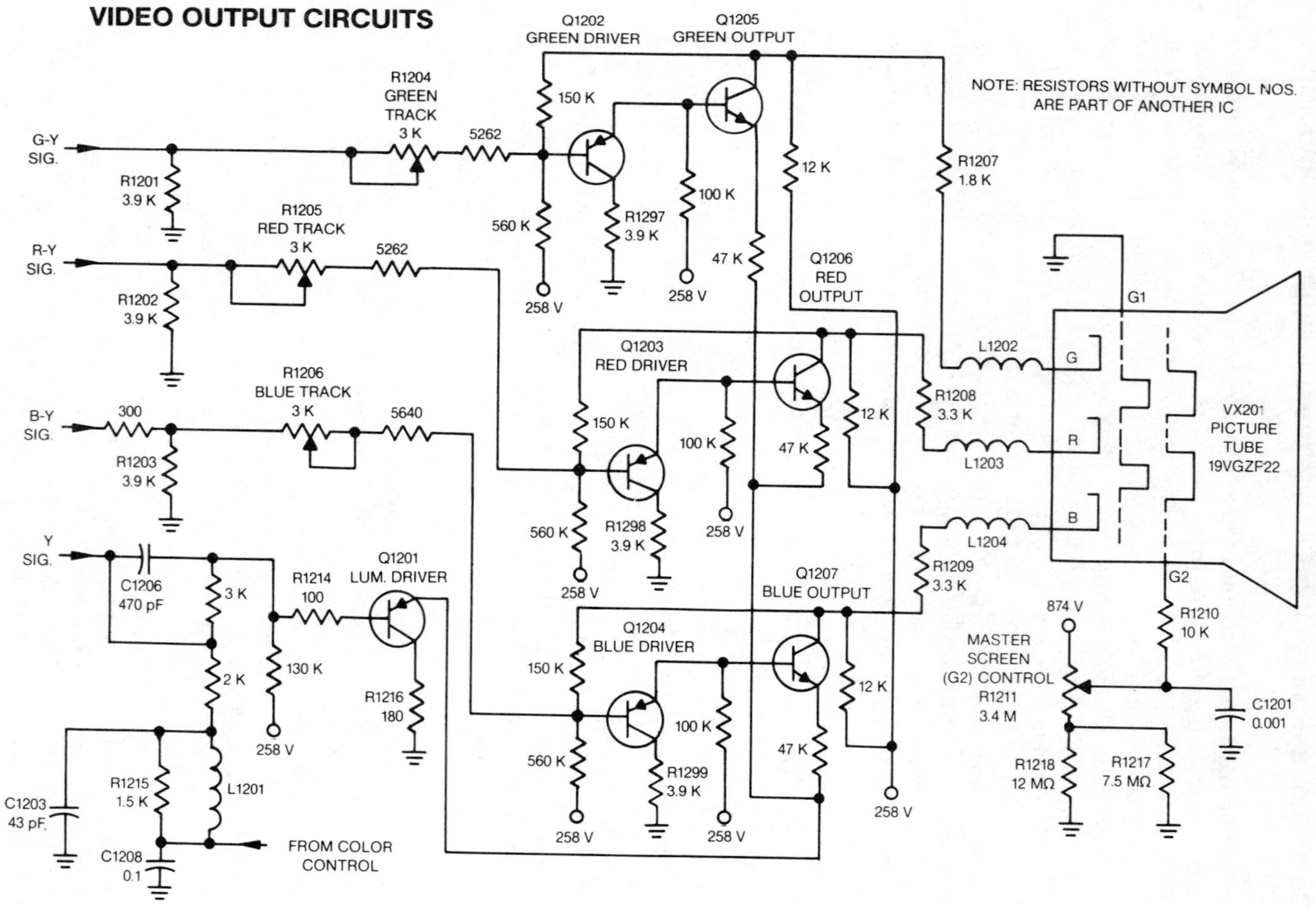
VIDEO OUTPUT CIRCUITS
Q1202 GREEN DRIVER
Q1205 GREEN OUTPUT
NOTE: RESISTORS WITHOUT SYMBOL NOS. ARE PART OF ANOTHER IC
G-Y SIG.
R1201 3.9 K
R1204 GREEN TRACK 3 K
5262
150 K
560 K
258 V
R1297 3.9 K
100 K
12 K
R1207 1.8 K
47 K
R-Y SIG.
R1202 3.9 K
R1205 RED TRACK 3 K
Q1206 RED OUTPUT
Q1203 RED DRIVER
L1202
G
R1208 3.3 K
L1203
R
B
L1204
R1209 3.3 K
G1
G2
VX201 PICTURE TUBE 19VGZF22
R1298 3.9 K
B-Y SIG.
300
R1203 3.9 K
R1206 BLUE TRACK 3 K
5640
Y SIG.
C1206 470 pF
3 K
2 K
R1214 100
130 K
Q1201 LUM. DRIVER
R1216 180
Q1207 BLUE OUTPUT
Q1204 BLUE DRIVER
R1299 3.9 K
874 V
MASTER SCREEN (G2) CONTROL R1211 3.4 M
R1210 10 K
C1201 0.001
R1218 12 MΩ
R1217 7.5 MΩ
C1203 43 pF.
R1215 1.5 K
L1201
C1208 0.1
FROM COLOR CONTROL

Low-Level Luminance Circuits

The low-level luminance circuits provide for video amplification of the luminance signal and the establishment of the necessary black-level control and chrominance-level control. All of these functions occur in a single IC in the Zenith model 19HC55, so this function will be discussed primarily in block diagram form, although a schematic diagram is included for reference.

As shown in the schematic diagram on page 4–164, the detected video is applied to the low-level luminance circuits via a delay line, described previously. The video to the chrominance circuits is taken from the video detector before the delay line. IC901, shown in the schematic, basically provides for video amplification of the luminance signal. The video signals are dc coupled, so it is necessary to establish a fixed, black level for the video. This is accomplished by clamping the back porch of the video horizontal-blanking pulse to a known level during each blanking period. In addition, the chroma signal (color reference) is brought into IC901 for adjustment of its gain consistent with the video gain (contrast) adjustment to ensure good tracking of the luminance and chrominance signals.

As also shown in the schematic diagram, the low-level luminance circuit is basically a single active IC (IC901) and an associated network of resistors packaged together as IC902 (resistors with values only). As illustrated, the video input is taken from the video detector and applied to the delay line (L208) (about 1.3 μsec) and also to the color demodulator. The delay line output is coupled to the IC via CR901, which clips negative-going noise pulses from the signal. C906 and L901 form a 3.58-MHz trap to recover the color subcarrier signal from the luminance signal.

The low-level luminance circuits are shown in block diagram form on page 4–165. IC901 contains the dual variable-gain video amplifiers operated by the contrast control, the gated clamp, and the output video amplifier. Not shown is a circuit that detects the vertical sawtooth. If the vertical sweep should fail, the CRT is biased off by a dc level applied to the output video amplifier that is dc coupled through the video output to the CRT cathodes. Since loss of horizontal sweep also causes loss of high voltage to the CRT, automatic protection against horizontal sweep failure is provided.

As shown, the video is applied to a variable-gain-controlled amplifier to provide video gain control (contrast control). The same type of variable-gain amplifier is also used to amplify the chrominance signal so that its gain is also controlled simultaneously. This is necessary because the chrominance data are taken before this gain control.

LOW-LEVEL LUMINANCE CIRCUITS

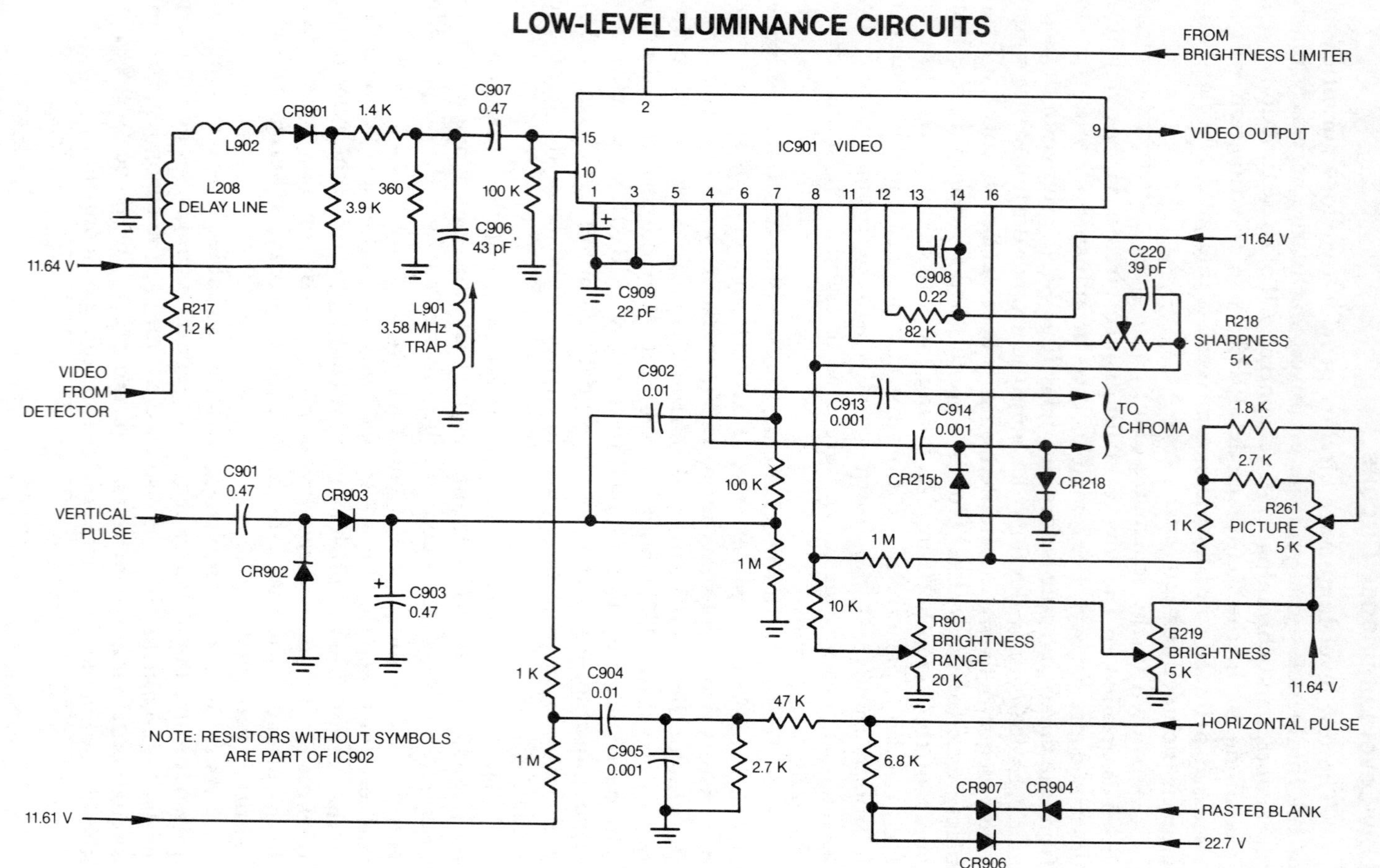

Low-Level Luminance Circuits (continued)

In many TV sets, a single control is used before the chrominance pickoff point. The output of the video amplifier has a dc component of the signal that depends on scene brightness. This variable dc would cause variation in the average CRT bias. To prevent this, the video signal is clamped to a fixed voltage (dc) by the gated clamp that establishes a fixed reference level. In this case, the clamp is driven by the retrace pulse from the horizontal sweep. This horizontal pulse is shaped and delayed in an RC network and used to drive an electronic diode switch that closes for the back porch interval of the horizontal-blanking interval. Since this level is fixed at the transmitter to represent the black level, all the video is referenced to this fixed level, as shown. The clamp output is fed to the video amplifier, as shown. Brightness is controlled by varying the dc level of the video output stage.

We will now proceed to a study of chrominance-processing circuits.

BLOCK DIAGRAM OF IC901

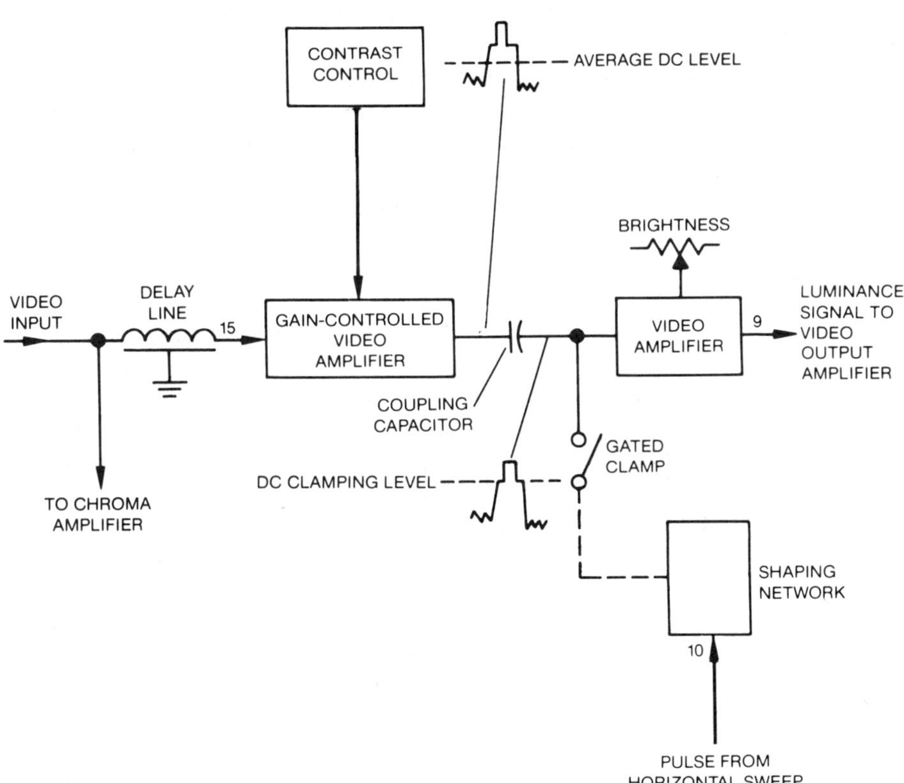

Chrominance-Processing Circuits

The last major element required to provide color TV reception is the recovery of the chrominance information. To do this requires that the 3.58-MHz color burst be separated and used to phase lock a local 3.58-MHz CW reference, and that this reference be used in conjunction with the chrominance video signal to obtain E_I and E_Q from which the color-difference signals R–Y, G–Y, and B–Y can be obtained. Since most modern color TV receivers use ICs, the overall IC circuits will be shown, followed by a block diagram description. It should be noted that many configurations of these circuits are used, however, the basic elements and functions described above are always present.

As shown schematically on the following two pages, the undelayed video input is applied to IC1001, which contains the color demodulators for recovery of the B–Y, R–Y, and G–Y signals. The video is coupled into IC1001 by L1002 and associated capacitances to eliminate low-frequency luminance signals. The phase reference is regenerated in IC1002. The horizontal blanking pulse is applied to IC1002 to gate the 3.58-MHz reference burst into the 3.58-MHz oscillator control. In the circuit shown, a 3.58-MHz crystal oscillator (CR1001) is used with IC1002 to provide a reference signal that is phase locked to the color burst. The phase-locked 3.58-MHz signal from IC1002 is then used in the phase detector/demodulator and matrix of IC1001 to provide the chrominance output signals. The E_I and E_Q signals recovered from the phase detector/demodulator are, in conjunction with the appropriate matrix, all contained in IC1001, so the E_I and E_Q signals are not a direct output.

The chrominance-processing circuits are shown in block diagram form on page 4–169. The video input is filtered to remove the low-frequency luminance components and applied to two phase detectors. The video input is also brought to a color-burst gate operated by the horizontal-blanking signal to gate the 8 cycles of 3.58 MHz through. The gated color burst is applied to a phase detector along with the output of the 3.58-MHz oscillator. The output of the phase detector is fed back to the 3.58-MHz reference oscillator circuit in such a way as to lock the phase of the 3.58-MHz reference oscillator to the color burst. The synchronized color-reference signal is then used to gate the video phase detectors. As shown, a 90-degree phase shift is included in one line to provide for demodulation of E_Q. The E_I output comes from the other phase detector, in which the 3.58 MHz signal has no phase shift.

The recovered E_I and E_Q signals are applied to the matrix shown to provide the R–Y, B–Y, and G–Y outputs for the video-output amplifier. The color-killer circuit shown is simply a detector for the 3.58-MHz color burst. When the color burst is present, indicating that a color program is being transmitted, the color killer allows the color demodulator to function normally. When no burst is detected, indicating that a black-and-white program is being transmitted, the color killer inhibits the demodulator circuits, and the R–Y, G–Y, and B–Y outputs are zero.

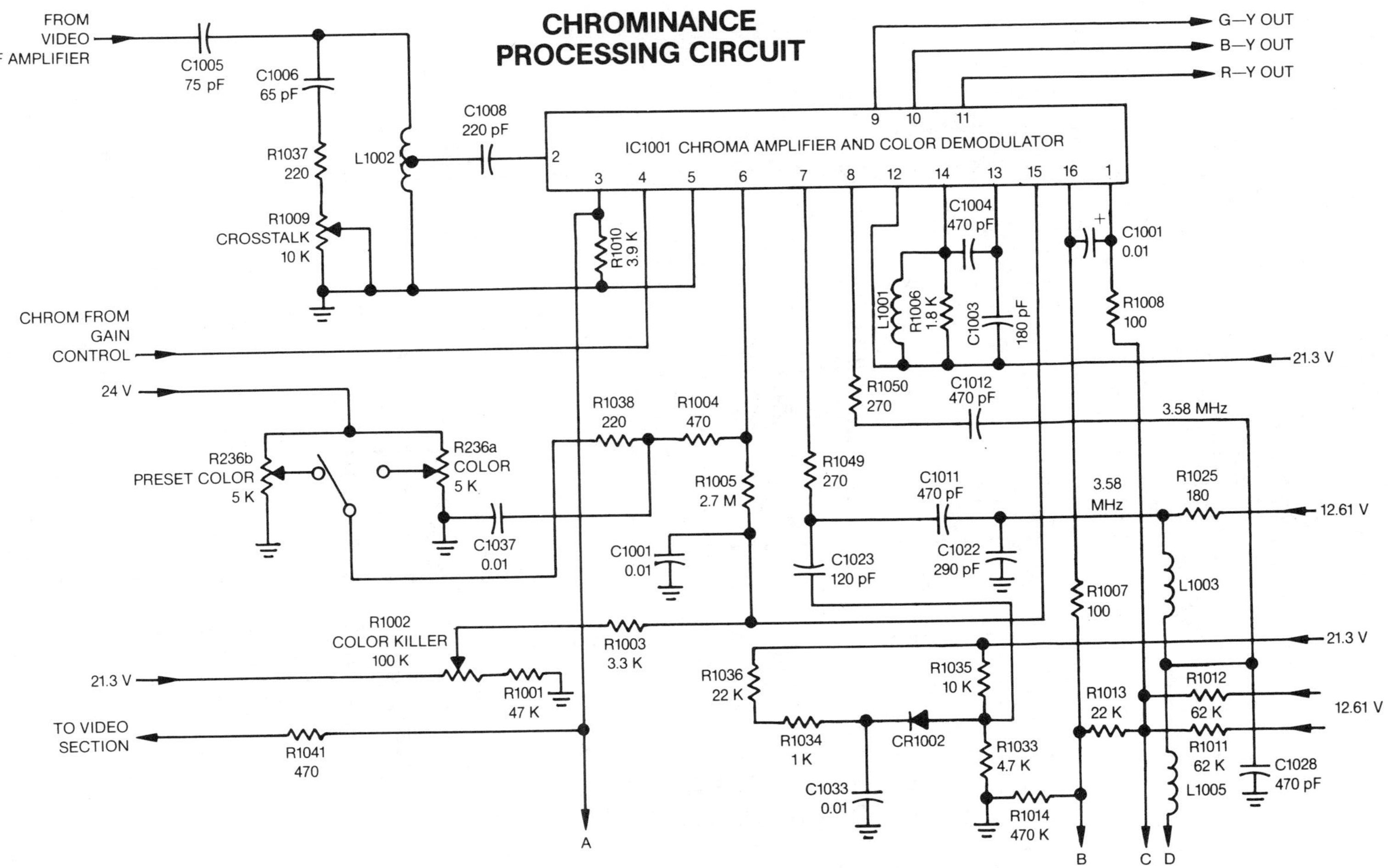
CHROMINANCE PROCESSING CIRCUIT
IC1001 CHROMA AMPLIFIER AND COLOR DEMODULATOR
FROM VIDEO IF AMPLIFIER
C1005 75 pF
C1006 65 pF
R1037 220
L1002
C1008 220 pF
R1009 CROSSTALK 10 K
R1010 3.9 K
CHROM FROM GAIN CONTROL
24 V
R236b PRESET COLOR 5 K
R236a COLOR 5 K
C1037 0.01
R1038 220
R1004 470
R1005 2.7 M
C1001 0.01
R1002 COLOR KILLER 100 K
R1003 3.3 K
21.3 V
R1001 47 K
TO VIDEO SECTION
R1041 470
A
G—Y OUT
B—Y OUT
R—Y OUT
C1004 470 pF
L1001
R1006 1.8 K
C1003 180 pF
C1001 0.01
R1008 100
21.3 V
R1050 270
C1012 470 pF
3.58 MHz
R1049 270
C1011 470 pF
3.58 MHz
R1025 180
12.61 V
C1023 120 pF
C1022 290 pF
R1007 100
L1003
21.3 V
R1036 22 K
R1035 10 K
R1013 22 K
R1012 62 K
12.61 V
R1034 1 K
CR1002
R1033 4.7 K
R1011 62 K
C1028 470 pF
C1033 0.01
L1005
R1014 470 K
B
C
D

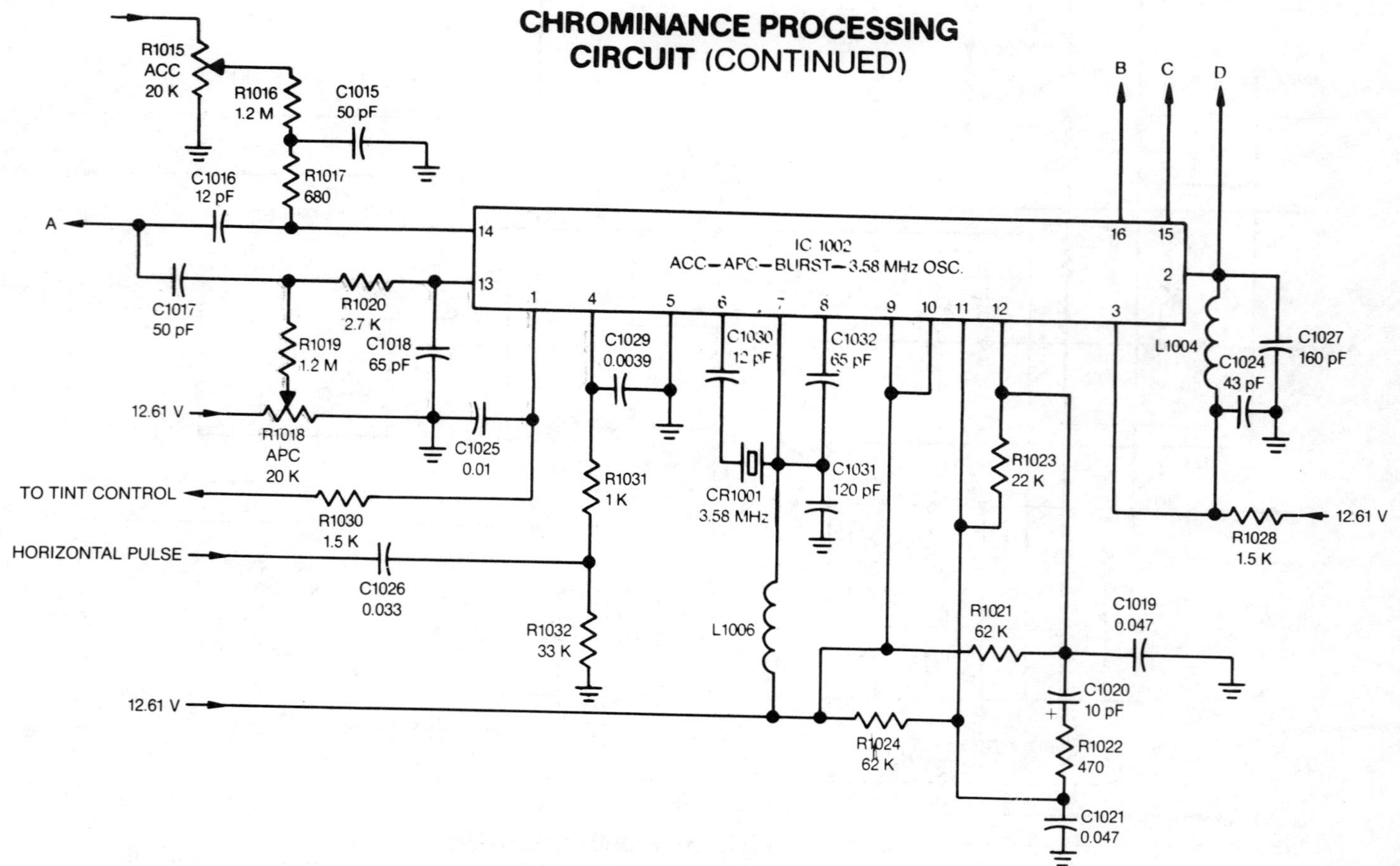
CHROMINANCE PROCESSING
CIRCUIT (CONTINUED)
IC 1002
ACC—AFC—BURST—3.58 MHz OSC.
B
C
D
A
12.61 V
C1027
160 pF
C1024
43 pF
L1004
R1028
1.5 K
C1019
0.047
C1020
10 pF
R1022
470
C1021
0.047
R1023
22 K
R1021
62 K
R1024
62 K
C1032
65 pF
C1031
120 pF
C1030
12 pF
CR1001
3.58 MHz
L1006
C1029
0.0039
R1031
1 K
R1032
33 K
C1025
0.01
C1018
65 pF
R1020
2.7 K
R1019
1.2 M
R1018
APC
20 K
R1030
1.5 K
C1026
0.033
C1015
50 pF
R1017
680
R1016
1.2 M
C1016
12 pF
R1015
ACC
20 K
C1017
50 pF
TO TINT CONTROL
HORIZONTAL PULSE

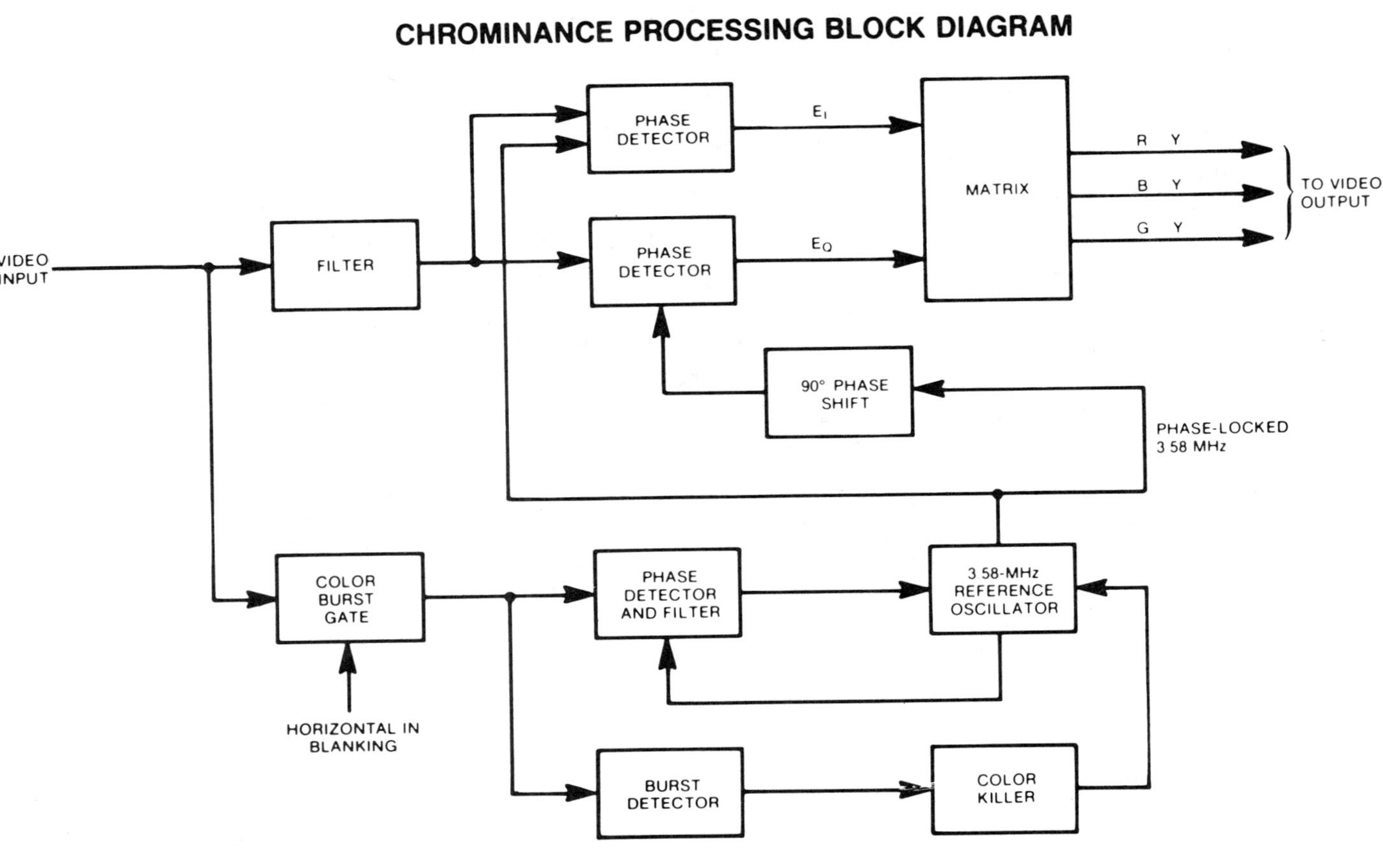
CHROMINANCE PROCESSING BLOCK DIAGRAM
VIDEO INPUT
FILTER
PHASE DETECTOR
PHASE DETECTOR
E_I
E_Q
MATRIX
R Y
B Y
G Y
TO VIDEO OUTPUT
90° PHASE SHIFT
PHASE-LOCKED 3 58 MHz
COLOR BURST GATE
HORIZONTAL IN BLANKING
PHASE DETECTOR AND FILTER
3 58-MHz REFERENCE OSCILLATOR
BURST DETECTOR
COLOR KILLER

Chrominance-Processing Circuits (continued)

The phase detector is at the heart of the chrominance-signal processing. You will recall from your study of phase detectors in Volume 3, that the phase detector provides an output signal that is proportional to the phase difference between two signal inputs. If the signals are in phase, maximum output is produced. If the signals are exactly out of phase, maximum output in the other direction is produced. If the signals are in quadrature (90-degree phase), no output is produced. The phase detector can be considered a switch operated by the reference signal.

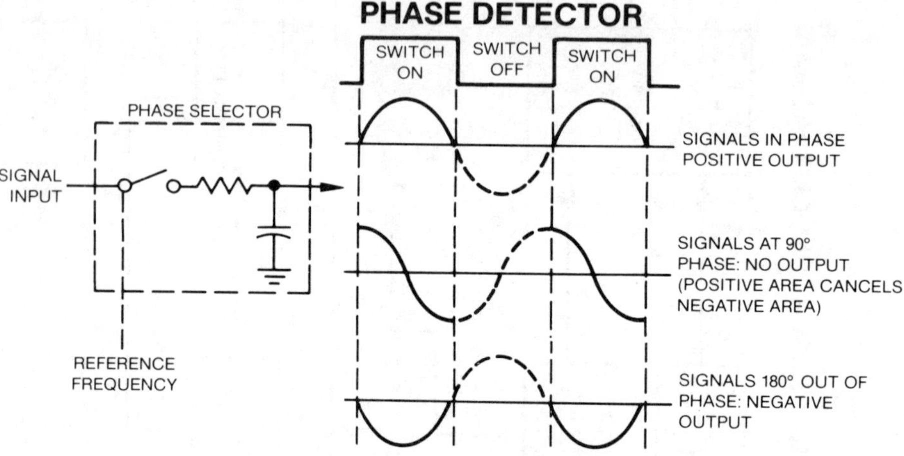

Consider first the operation of the phase detector in the phase-locked 3.58-MHz reference-oscillator circuit. Here, the phase of the color burst from the incoming video is compared to the phase of the 3.58-MHz reference oscillator. If a phase error exists, a voltage output from the phase detector is produced. This voltage can be filtered and applied to a tuning element such as a varactor across the crystal to adjust its frequency slightly. Since a feedback loop is involved, the circuit will adjust itself so that the two signals are in a fixed relative phase—actually, they will be in quadrature (90 degrees), since this is the null point of the phase-detector characteristic. Thus, the output of the reference oscillator is in quadrature and can be used directly for the E_Q demodulator. In a practical circuit, the 90-degree shift would be in the E_I phase-detector line.

Now that a stable phase-locked reference is available, it is a simple matter to apply it to two-phase detectors to recover the I and Q signals. The tint control on a TV receiver is actually a phase trimmer that shifts the phase of the reference (or one of the components of the reference) to compensate for any undesired phase shift in the color circuits. Thus, a phase-detected output is obtained which is aligned with the transmitted chrominance signal. The color (or hue) control adjusts the level of the chrominance-difference signals (R–Y, B–Y, G–Y). When turned completely down, only the luminance signal drives the output (black and white). As the level is increased, the intensity of the color image is increased.

Beam Convergence Control—Color Purity

As stated earlier, it is necessary to assure that the red, green, and blue beams strike only their corresponding color phosphor dots. If the illuminated dots on the picture tube screen do not *exactly match* those being scanned *at that instant* in the transmitter camera tube, the sharpness of original image will be lost, and distortion in color will take place.

The cause of the convergence problem is the fact that the distance between the electron guns and the screen increases as the beams are scanned away from the center of the screen. In addition, the beams do not occupy the same position in the deflection field. Because the distance increases in a regular predictable manner and the relative position of the beams in the deflection field is known and fixed, a correction can be made. All color TV receivers contain special *convergence-control circuits* that aim the electron beams at the same triad of dots and make corrections for the curvature of the screen and deflection.

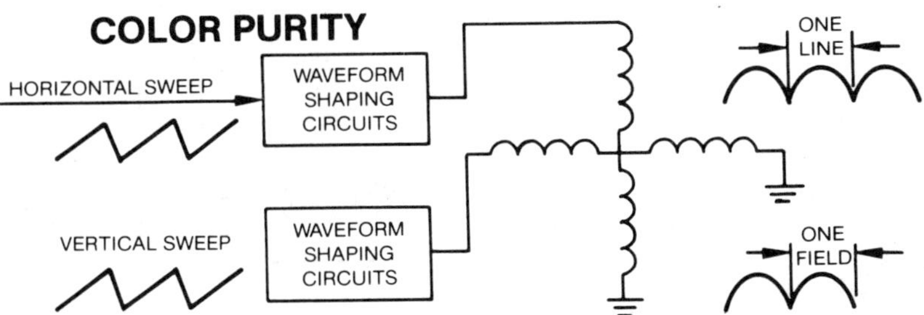

The key to making this correction is the fact that the amount of deflection of the electron beams away from the center is controlled by the magnitude of the electric current through the vertical- and horizontal-deflection coils. Since these currents determine exactly the angle that the beams will deflect, they can be used to make the precise correction required to make all the beams converge on the same triad at all times by driving them with the proper waveform.

The correction is made by placing a horseshoe-shaped electromagnet on the neck of the CRT, in back of the deflection coil. Vertical and horizontal coils are mounted on each magnet. Applied to each coil is a variable-signal waveform that is in step with the vertical- or horizontal-sweep signal. The magnitude of the magnetic fields produced by these coils can be made proportional to beam deflection, so that adjustment of the waveform and the current through these coils can be made to have the three beams always *converge* on the *same* triad of dots.

The horizontal and vertical sweeps are shaped by inductors, capacitors, and resistors to produce the desired waveforms for application to the convergence coils. Since the distances to all parts of the screen from the centers of deflection of the beam are not constant, some defocusing will result at the edges of the picture, if the center is focused. This defocusing will cause parts of a beam to leak through adjacent openings in the shadow mask, which is undesirable. To compensate for this, a correction voltage is often added to the focus voltage.

Features of Modern Color TV Receivers

The color TV receiver described earlier is typical and will allow you to understand how any color TV receiver operates. You will be able to understand circuit function, even though the circuit details may vary. There are, however, many variations of remote-control features and tuners. Most remote controls are *ultrasonic devices* that actuate an ultrasonic microphone in the receiver. Differing tones (or differing codes) determine what function is to be performed. For example, audio volume, contrast, or tint are often simply motor-driven controls that are actuated by the ultrasonic signals, so that the longer the appropriate button is depressed, the greater (or lower) the volume or other function becomes.

COLOR TV RECEIVERS

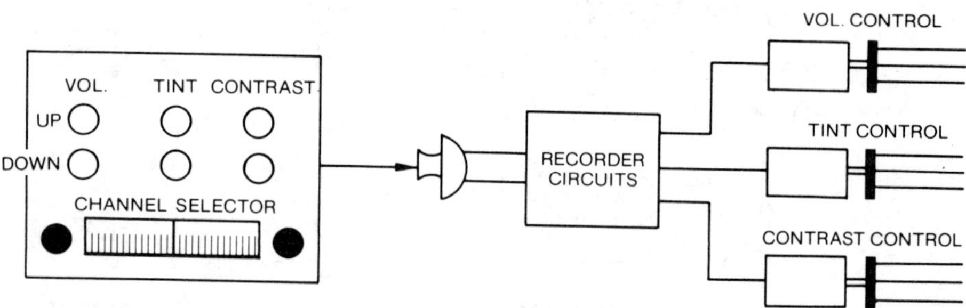

In simplest form, remote tuning is accomplished by having a motor, actuated by the remote control, that sequences the tuner through the channels, with the motor stopped when the remote control signal is removed by the viewer when the desired channel is reached. Usually no provision is made for fine tuning, with the assumption that the memory tuner has been preset earlier. In general, simple remote tuners do not provide access to the UHF channels.

Some TV sets provide for access of 10–18 or more channels that can be preset. Thus, the VHF and UHF channels available in a given area can be pretuned. It is then a simple matter to access these by a remote control. In some cases, this preset is mechanical, but in others it is done by presetting a dc voltage for each channel that operates varactor tuners, and then the only thing necessary to do is to select the proper voltage for a given channel and apply this to the tuning varactors. This approach provides a very flexible approach to remote tuning of both VHF and UHF channels. It is also possible to trim this applied voltage remotely, providing for simplified remote fine tuning. Because of difficulty with seeing the channel number on a TV set from a distance, some TV sets are designed to flash on the TV screen the channel number to which the set is tuned.

Features of Modern Color TV Receivers (continued)

The availability of low-cost microprocessors has greatly enhanced the flexibility of TV set control, both locally and remotely. For example, a microprocessor is used by the Zenith Radio Corporation in some of their deluxe-model TV receivers. This microprocessor is used in conjunction with a phase-locked-loop frequency synthesizer to provide for flexible remote electronic tuning of all TV channels as well as other special channels for CATV. While understanding of these digital circuits must await your study of digital systems in Volume 5, some insight can be provided here on system operation, since you already understand how phase-locked loops operate.

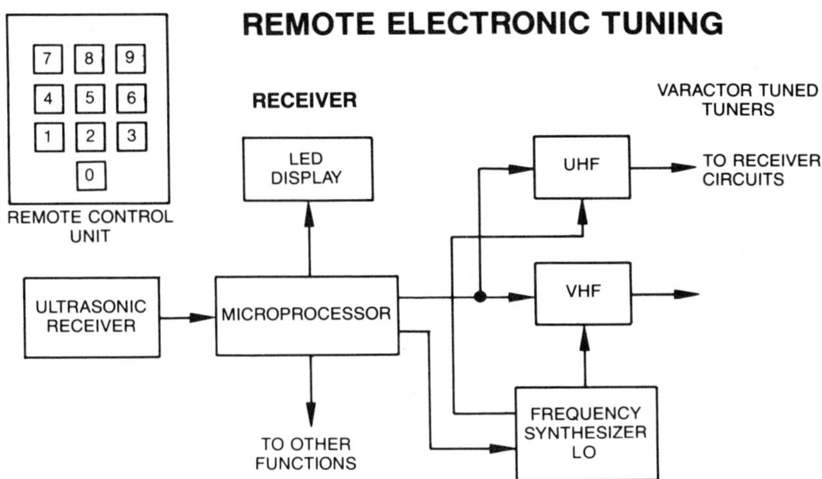

In operation, the input to the microprocessor is supplied as digital signals from the remote control. These are picked up by the ultrasonic receiver and fed to the microprocessor. The microprocessor interprets the received signal and generates voltages that tune the TV receiver tuners (RF, mixer, LO) to the proper channel. A frequency synthesizer and phase-locked loop are used to automatically fine tune the LO to the exact frequency for the channel selected. In addition, a digital signal is produced to operate a LED (light-emitting-diode) display (such as on a digital watch or clock radio) to indicate the channel selected. The channel-selector keyboard on the remote unit (or local-control unit) looks very much like a calculator keyboard. To select, for example, channel 21 requires only that the 2 and 1 pushbuttons be depressed, followed by ENTER. The receiver will automatically tune to channel 21, lock to the correct frequency for exact tuning, and display 21 on the LED channel display. The microprocessor is also used to route the other remote-control functions, such as volume, to the proper actuating circuits. It should be stressed that while some convenience feature like remote control or automatic fine tuning may vary widely from receiver to receiver, *the fundamental ideas are the same.*

Experiment/Application—TV Receivers

A lot can be learned by exploring a TV receiver with an oscilloscope and probe. If you set up a receiver and tune it to a channel that is functional, then all of the signals found in a TV receiver will be present. Thus, it will be easy to see TV signals in action. *CAUTION: Do NOT touch the high voltage portion of the receiver.*

EXPLORING A TV RECEIVER WITH OSCILLOSCOPE AND PROBE

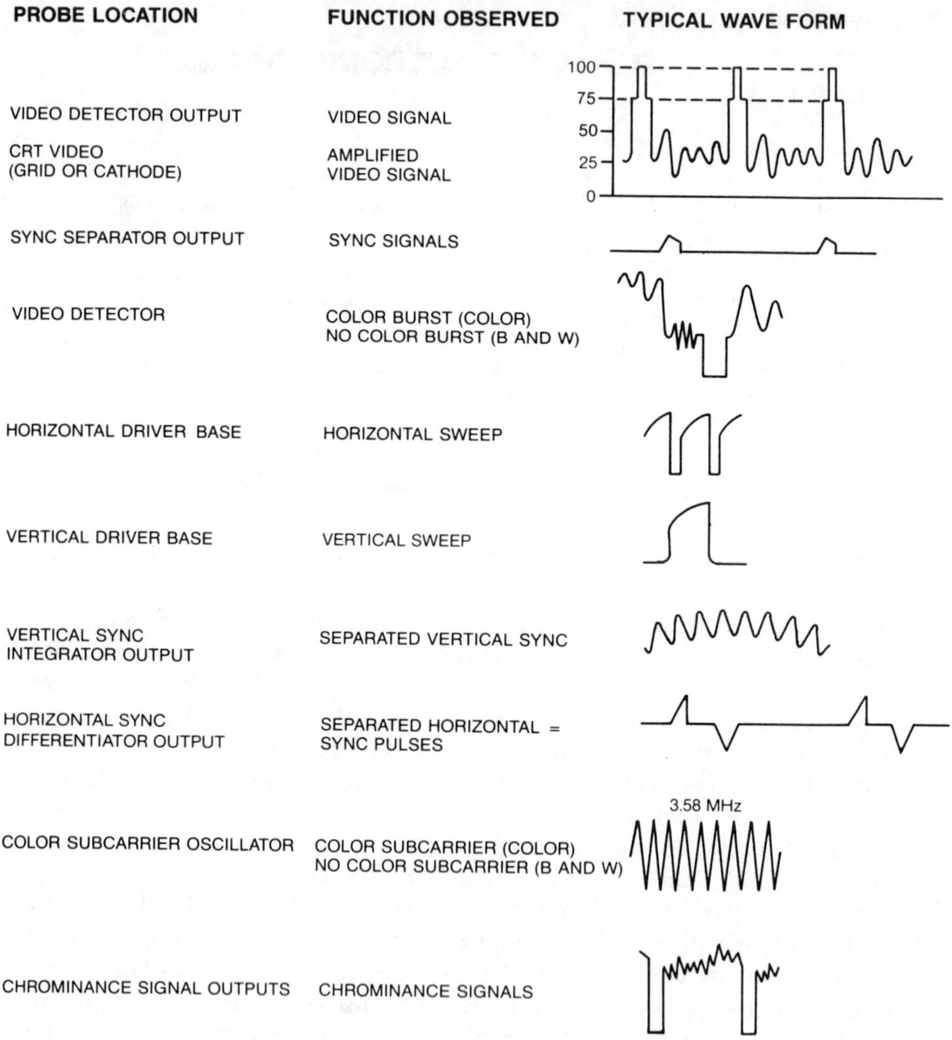

You can explore other portions of the video/audio section of the TV receiver, and it is of interest to examine the above waveforms with the TV set in sync and out of sync.

Video Tape Recording—Broadcast Type

In the early days of TV, recording of programs was done on film—either black and white or in color. Video recording on tape was impractical then, because of the signal bandwidth requirements. Improvements in recording technology have made it possible to record video *directly on magnetic tape*, and now such devices are commonplace and relatively inexpensive. To make such a recording on tape by the same techniques as for audio recording would require high tape speeds —a great deal of tape for a short program. The breakthrough that made TV recording practical was the development of the *rotating recording head.*

In these systems, the tape moves horizontally, as for a conventional recorder, but the head rotates, moving the head across the tape at high speed and laying down a strip of recorded data with each pass of the head over the tape. This strip of recorded data is *perpendicular* to the direction of motion of the tape. Typically, four heads are mounted on the rotating head assembly so that as one head leaves the tape, another moves onto the tape to continue recording.

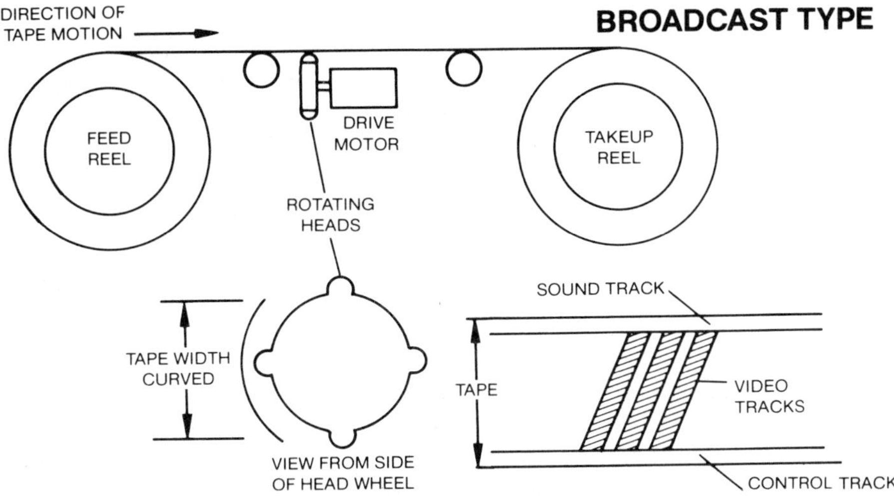

As shown above, the video recording is laid down in strips by the high-speed rotating head. The tape is curved into quarter circles so that the rotating head is in close contact with the tape during its traverse. A sound track and a control track are laid down in conventional fashion. The sound track is used to record the accompanying audio. The control track is used to carefully regulate and control the tape motion and speed. Typically, the tape speed is 15 inches (38.1 cm) per second. The head rotates at 14,400 rpm, and the width of each recorded strip is about 0.01 inches (0.025 cm). A 2-inch (5.08-cm) diameter head rotating at 14,400 rpm produces an equivalent tape speed of 1,500 inches (3,810 cm) per second. This allows for the recording of the high video frequencies necessary for high-fidelity TV tape recording. To achieve the best fidelity, the TV signal is used to frequency modulate a carrier.

Video Tape Recording—Portable Type

The portable and home type video tape recorders use a type of scan called a *helical scan*. In these recorders, the tape is wrapped partially around a fixed capstan that has the rotating recording head placed at an angle. The heads scan the tape at an angle. Because of this, each individual scan of the tape head can be made much longer, as shown below.

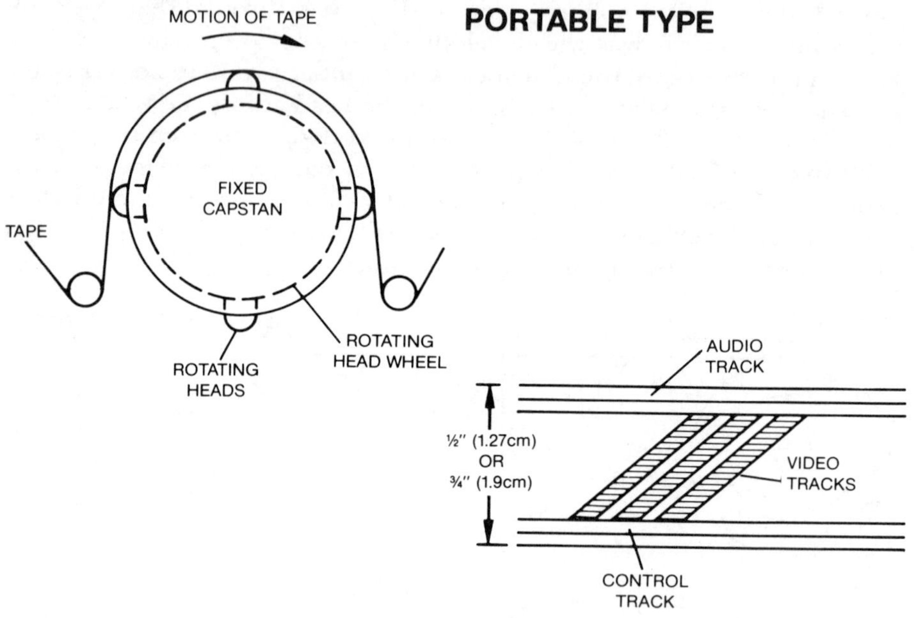

Depending on the capstan and rotating head wheel design, either two or four heads are used. In addition, the tape may be wrapped halfway around the capstan, as shown, or almost entirely around the capstan. The heads are actually rotated in a slot in the fixed capstan so that they can make contact with the tape for recording.

In other respects, the home or portable tape recorder is similar to the larger video tape recorders described on the previous page. Unlike the broadcast-type video tape recorder, there are no real standards and several types of machines are available with different characteristics; unfortunately, they are incompatible. In addition to ½-inch (1.27-cm) and ¾-inch (1.9-cm) tape models, open-reel and cassette designs are available, and, in addition, the recording technique varies somewhat. In a modern home tape recorder/player, up to 6 hours of recording can be obtained with a single tape cassette, with surprisingly good quality. Since an entire frame is recorded on one traverse of the head (half or full circle), it is possible to stop motion and look at a single frame by stopping the tape motion while the head continues to rotate.

Video Disk Players

Video disk recording for home entertainment and other purposes is becoming increasingly common. The reason for this development is primarily that phonograph disks continue to be the most popular for audio recording. They are easy to handle and change, and they are easy to mass produce—leading to low cost. Because of the complexity of recording and manufacture, video disk systems are used for playback only. Two basic systems have been devised. One system uses a laser beam to read the data from the disk. The other uses an extremely fine diamond stylus that acts as a capacitor plate to read the data from the disk. Video disks look like an ordinary phonograph record. Programs of all kinds are available in both color or black and white. Both systems produce a video signal fully compatible with current TV standards.

The laser beam player was developed by the North American Phillips Company and is marketed under the trade name of *Magnavision*®. In this system, the disk is rotated at a speed of 1,800 rpm. As with video tape, the signals are recorded and played back using FM. The sound channels (stereo) are subcarriers. The FM carrier is at 8.1 MHz, modulated by the video signal, including color signals. The sound subcarriers are at 2.1 and 2.3 MHz. The data are recorded as a series of microscopic pits of variable length, with a width of 0.4 μm (μm = micrometer = one-millionth of a meter). The grooves are spaced about 1.6 μm apart.

VIDEO DISK SYSTEM

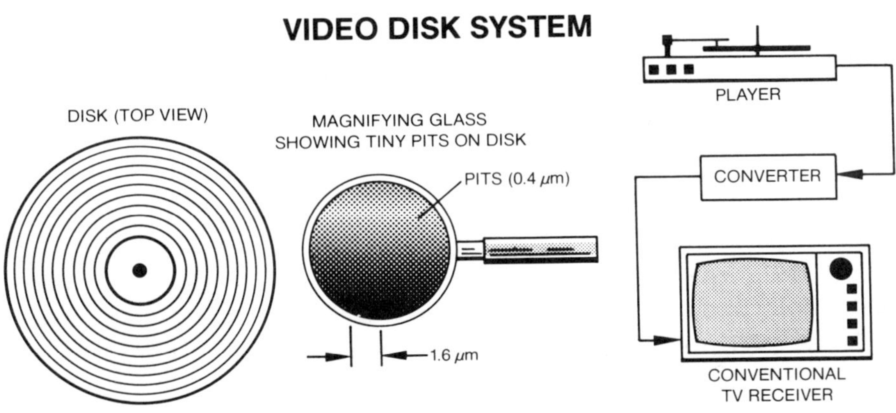

The disk is read with a laser beam that is split into three beams. The two outer beams are used to make the center laser beam track the recorded data. The output signal is converted to baseband video and audio and can be used directly or to modulate carriers, just as in a conventional TV transmitter, except for the very low power output (a few milliwatts). The carrier is set for an unused TV channel on a conventional TV receiver and is received just like any TV broadcast, so the player can be used with any TV receiver. The optical system used to read the disk is described on the next page.

Video Disk Players (continued)

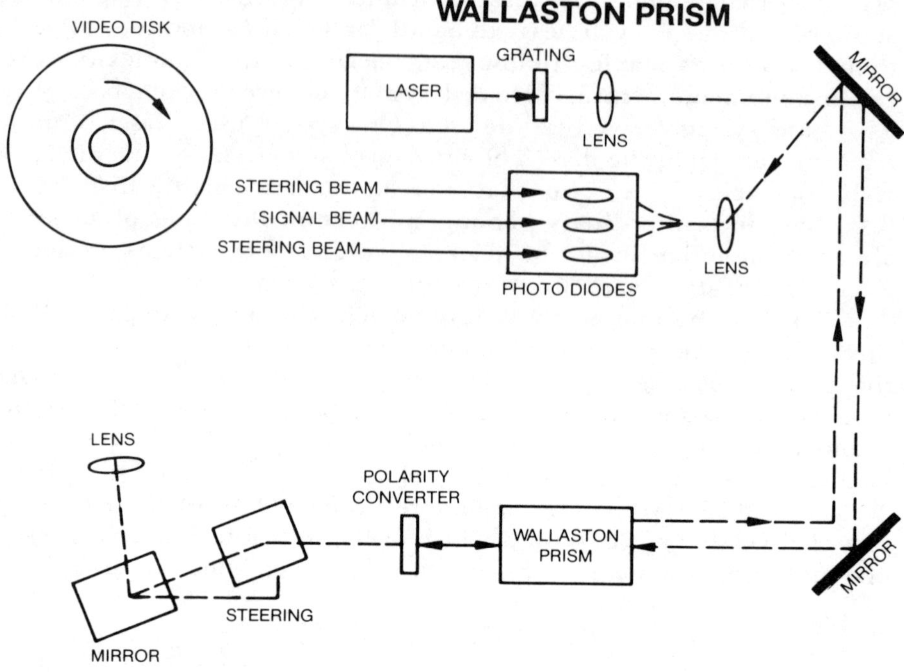

For the laser disk system, a laser beam is split into three beams by a grating, and these beams are focused by a lens onto a fixed mirror system, as shown. The light beams are directed through a *Wallaston prism* that bends the light differently for different polarizations—that is, *vertically* polarized light is bent in a direction *opposite* to that of *horizontally* polarized light. The light from the laser coming from the prism is vertically polarized, and the polarity converter converts it to a circularly polarized light that is directed via the steering mirrors through a lens onto the disk. The light reflected from the pits in the disk is less intense than the light reflected from the area between the pits, producing an intensely modulated reflected beam. This return beam is passed back through the steering mirrors to the polarity converter and prism. Because the beam is going in the opposite direction, it is deflected at the output of the prism in the opposite direction from the incoming laser beam; thus the laser and return beams are separated. The return beam is reflected by the fixed mirrors onto a set of photo diodes that produce the necessary output signals from the center beam and the beam-alignment data from the two outer beams. The video and audio output signals are generated by demodulation of the FM signal.

Video Disk Players (continued)

The RCA video disk system operates differently. The disk rotates at 450 rpm and is read by a diamond stylus. The stylus follows the groove just as with a conventional phonograph record, so no special circuitry is necessary to hold the stylus aligned. The information is coded as a series of transverse slots of varying width in the bottom of the groove. The diamond stylus has a thin metal electrode plated on its rear surface. The record itself is made from a conductive plastic or a thin plating of conductive material. These act as the plates of a capacitor, and the stylus reads the stored data by the *variations* of this *capacitance* as the record rotates.

DISK/STYLUS OPERATION

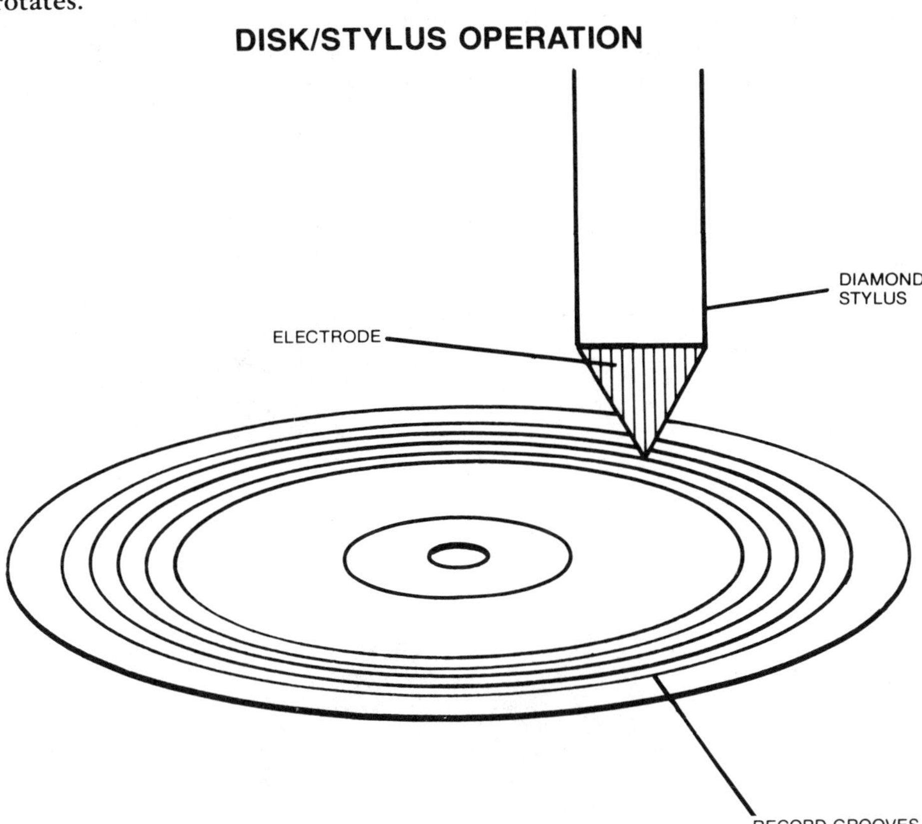

The recording technique, as noted previously, basically uses FM carriers. Separate carriers are used for the video and audio information; however, special subcarrier and encoding techniques allow for the use of a lower carrier frequency, which can lead to longer playing time per side than for optical techniques. While the lower rotating speed is an advantage that leads to a relatively low-cost system, special features such as stop motion and slow motion appear to be difficult to achieve.

TV Receiver as a Video Display Terminal

Now that you have learned all the aspects of the TV receiver when used as a TV receiver, we would like you to take another look but from a *different point of view*—that is, viewing the TV receiver as a *video display terminal* when used with a computerized information system.

As you will learn in Volume 5, the CRT terminals used with computers/microprocessors consist of the deflection and sweep circuits, the video portion, and the necessary power supplies. Additional hardware is used to provide the video and sync signals for the basic CRT display. These terminals have sweep characteristics very similar to those of a TV receiver—typically, a 60-Hz vertical field rate and about 500 lines. You also know that an interface is available which will take incoming video and convert it to an RF signal that is set for one of the TV channels. Thus, an incoming video signal is displayed on the TV receiver CRT.

TV RECEIVER AS A VIDEO DISPLAY TERMINAL

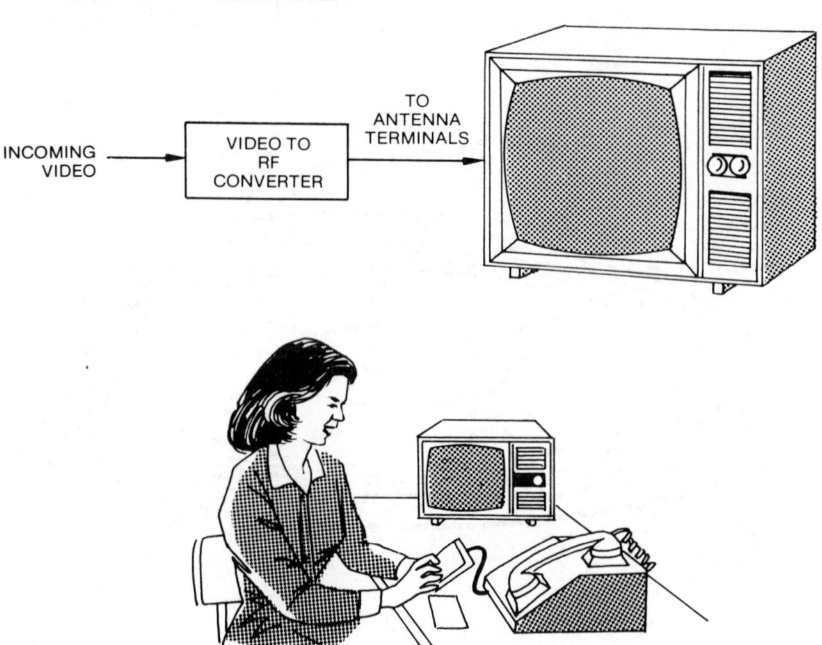

Digital data, a series of on-off signals that represent alphabetic or numeric characters, lines, or bars, etc., are really no different from TV information. Such digital data can easily be *formatted* (arranged) with the proper digital circuitry to provide a display of information on a regular TV receiver.

TV Receiver as a Video Display Terminal—Teletext

TV receivers can be (and are) widely used as *display terminals*. The necessary information from any source for display is stored in a memory device, and this memory device is scanned at the TV scanning rate to provide the necessary display. You will learn about these storage and interface devices when you study digital circuitry in Volume 5 of this series.

One of the new uses of TV (video) display terminals is in a *Teletext* mode. Teletext is data, usually alphanumeric, inserted into the TV transmission during the vertical (or horizontal) retrace interval. The data are displayed on the TV screen either alone or as a caption or separate information unrelated directly to the program material. Aside from providing simple captions, headlines, traffic information, supermarket specials, etc., it has obvious uses in providing subtitles for those with impaired hearing or providing translation of foreign languages. In addition, so-called *electronic home information services* are being developed which will use cable TV systems as a two-way link into textual and graphic material upon demand, and computer/microprocessor linkup as desired.

TELETEXT SYSTEM

The Teletext approach is the basic step in the use of TV for *home electronic retrieval of information*. A more elaborate system called *Viewtron* provides a much broader basis for *interactive operation* involving not only information retrieval and computer hookup but also access to a broadly developed *data base of information*. Future uses such as comparative shopping at home from supermarkets or other stores (using their existing price structure as stored in their computer systems for specific items) will be readily possible once these systems are in operation.

Now that we have completed our study of the TV receiver as a video display terminal we are ready to tackle the subject of troubleshooting receivers. It is important here to keep in mind the principles of troubleshooting that we have learned in the first three volumes.

Troubleshooting Receivers

Receiver troubleshooting is somewhat different than transmitter troubleshooting in that, as you might suspect, the signal-tracing procedure starts from the output end, as shown on the following page. As always, the first thing to make sure of is that there is primary power and that the power supplies are working properly. If this is so, it is then important to check the audio system. This is most easily done by injecting a low-level audio tone into the input of the first audio stage and tracing forward for the presence of the proper audio level if no signal or a weak or distorted signal appears at the output.

If the audio section is operating properly, the IF/detector system can be easily checked by loosely coupling an IF signal into the mixer collector. Most RF–IF signal generators have provisions for tone modulation, which is convenient. If the tone is heard at the receiver output and appears generally normal, then the local oscillator (LO) should be checked. No output from the LO will, of course, make the receiver totally inoperative. LO operation can be difficult to check unless a detector probe is available. Such a probe can be made as shown on the following page. A dc output indicates that the LO is functional, which means that the RF/mixer needs to be examined. This is most easily done by injecting a signal from an audio-tone modulated RF signal generator into the mixer input first and then, if this is working, directly into the RF input.

As with transmitters, it is extremely important *not to touch adjustments* until the receiver is functioning. Otherwise you could repair the defect but still not have a functioning receiver because of misalignment. This can lead you to believe that the receiver is still defective.

As with any electronic equipment, repair or troubleshooting is difficult without help in the form of schematic diagrams, voltage and signal levels at pertinent points, and layout diagrams. In addition, most equipment includes some relatively specialized alignment techniques. Thus, the use of manufacturers' data and alignment and troubleshooting procedures makes the job of troubleshooting much much easier than if you try it blind.

Generally, after repair it is wise to check the alignment of a receiver since misalignment can cause serious loss of sensitivity and selectivity. While this procedure differs for different receivers, the general approach will be described next; it can be used as a guide if no other data are available.

Now, study the following chart *carefully*, refer back to it if in doubt, and, above all, make sure that you have not left anything out in each step of the process.

The next section on receiver alignment will also help in the troubleshooting procedure.

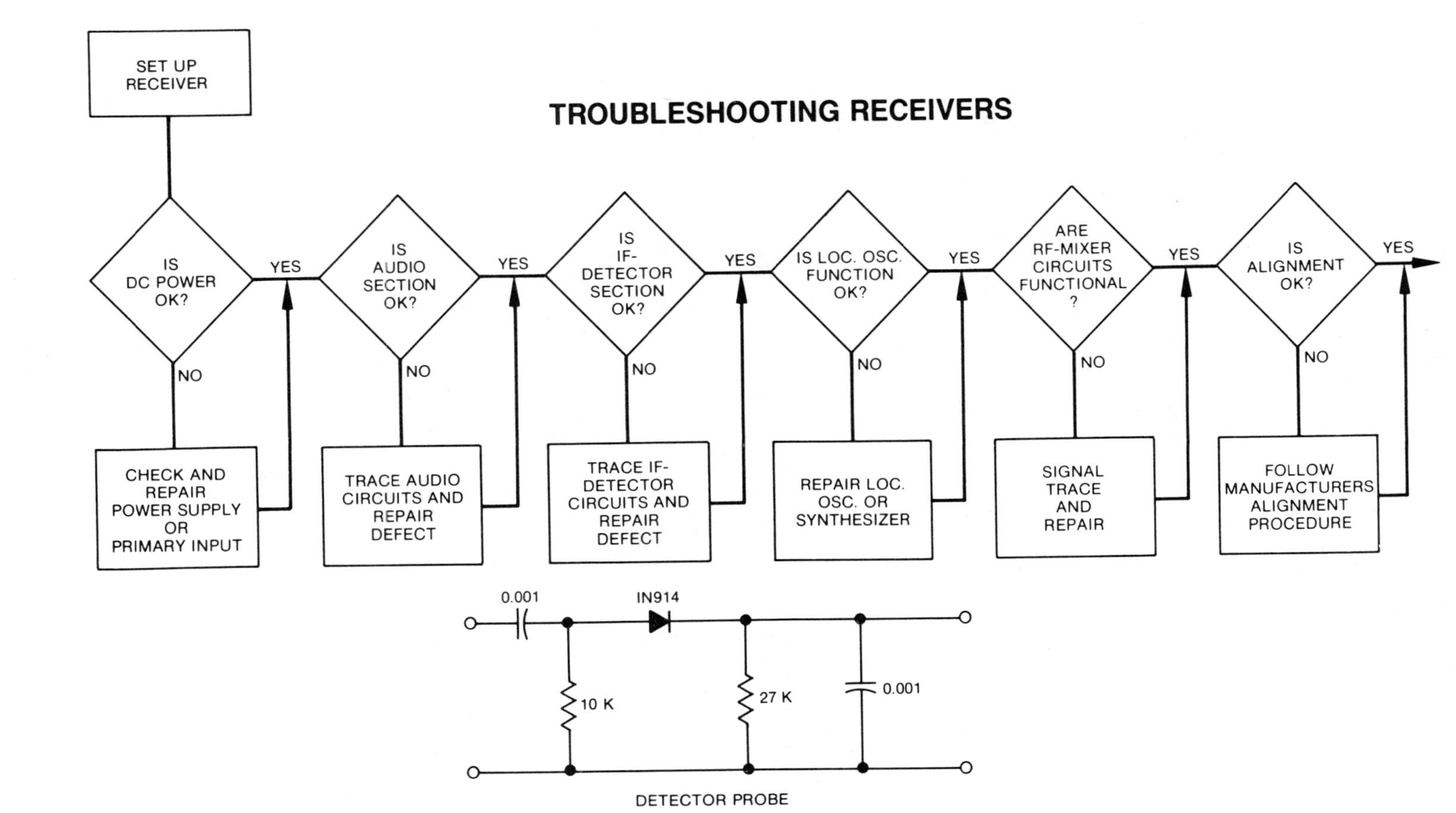
TROUBLESHOOTING RECEIVERS
SET UP RECEIVER
IS DC POWER OK?
YES
NO
CHECK AND REPAIR POWER SUPPLY OR PRIMARY INPUT
IS AUDIO SECTION OK?
YES
NO
TRACE AUDIO CIRCUITS AND REPAIR DEFECT
IS IF-DETECTOR SECTION OK?
YES
NO
TRACE IF-DETECTOR CIRCUITS AND REPAIR DEFECT
IS LOC. OSC. FUNCTION OK?
YES
NO
REPAIR LOC. OSC. OR SYNTHESIZER
ARE RF-MIXER CIRCUITS FUNCTIONAL ?
YES
NO
SIGNAL TRACE AND REPAIR
IS ALIGNMENT OK?
YES
NO
FOLLOW MANUFACTURERS ALIGNMENT PROCEDURE
0.001
IN914
10 K
27 K
0.001
DETECTOR PROBE

Receiver Alignment

Optimum performance of a superheterodyne receiver cannot be obtained unless it is properly tuned, or *aligned*. The process of alignment involves three things:

1. The IF amplifiers must be tuned to the intermediate frequency.
2. The tuned RF circuit(s) must be tuned to the frequency on the dial.
3. The local oscillator must be adjusted to give an output at each setting of the tuning dial that is above (or below) the RF by an amount equal to the IF.

The alignment procedure will be described by referring to an AM slug-tuned receiver, although an FM receiver is aligned in a similar way. The first step in the alignment is to tune the IF transformer to the IF. This is done by loosely coupling a signal from a signal generator set to the IF into the collector circuit for the mixer.

The antenna can be disconnected and a test oscillator coupled to the input through a small capacitor (27 pF). With the dial set at any convenient frequency, the test oscillator is adjusted to provide a signal at IF. The tuning-slug adjustments in the IF are then adjusted to provide a maximum output.

It should be noted that not all IF sections are aligned with every stage at the same frequency. Some are *stagger tuned*. As you have previously learned, this means that some IF stages are to be peaked at different frequencies, in order to broaden the overall response curve. The procedure then is to use a modulated signal from the generator and connect the probe to the detector output. Each IF stage is then peaked up at its own designated frequency (called for in service information). In the case of an FM receiver, with a limiter stage, the lowest possible signal input should be used, to avoid limiting action, or an RF probe can be used at a point just before the limiter.

The RF amplifier, mixer, and LO tuning are adjusted first at the high end of the band where the trimmer capacitors have maximum effect. Both the tuning dial and test oscillator are set at near the high end of the band, and then the trimmers are adjusted for maximum signal level at the output. Care must be taken to use a low-capacitance probe when measuring the RF signal in order to avoid detuning the RF amplifier output. Next, the signal generator and tuning dial are set to the low end of the band and the tuning-slug position is adjusted to provide maximum signal level. This tuning can change the high-frequency alignment. Therefore, the dial and generator are both tuned back to the high end, and the trimmers are readjusted for maximum output.

Alignment of the oscillator section is similar to the RF amplifier procedure with the trimmer being tuned first at the high end and the coil then tuned at the low end. Again, there will be an interaction between the effects of tuning, and the procedure should be repeated several times while tuning for a maximum at the IF output.

AM BROADCAST RECEIVER TUNING ELEMENTS

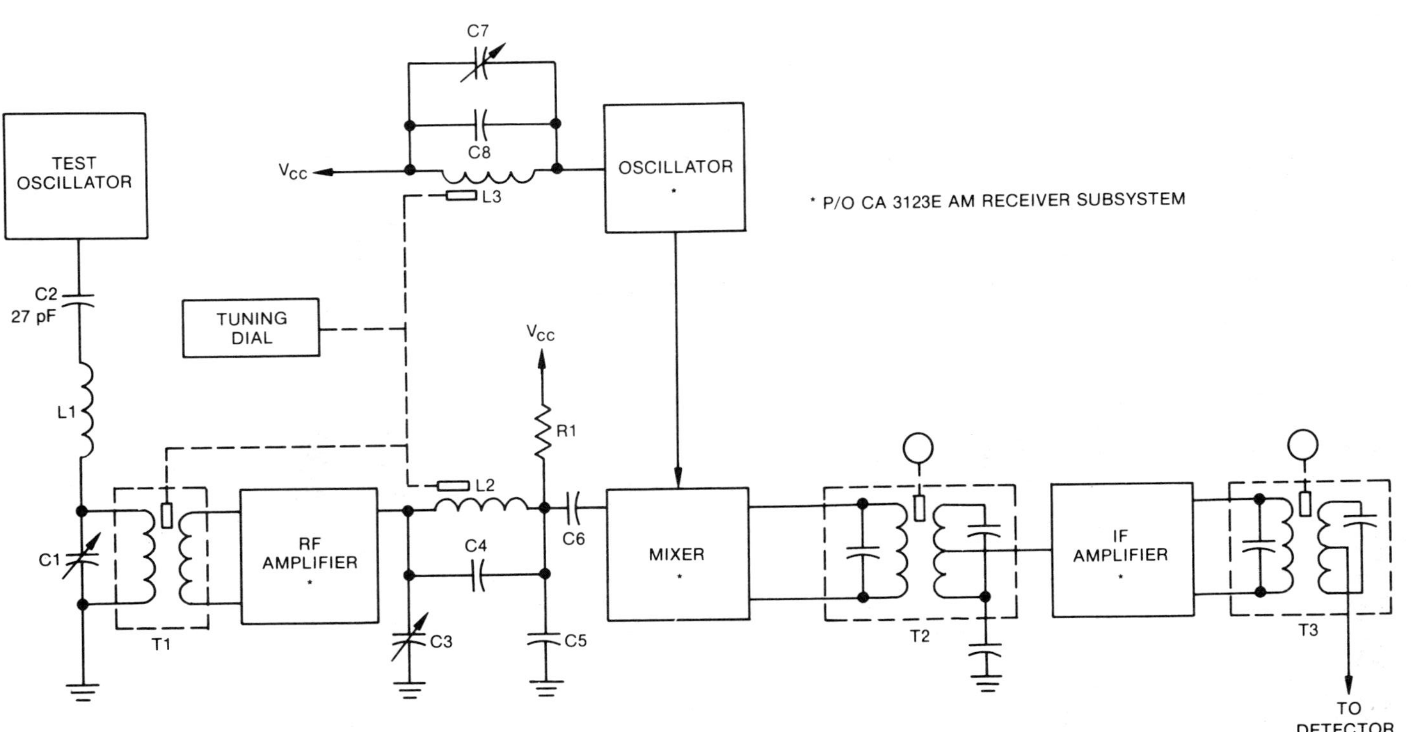

Review of Color TV Receivers and Video Recorders

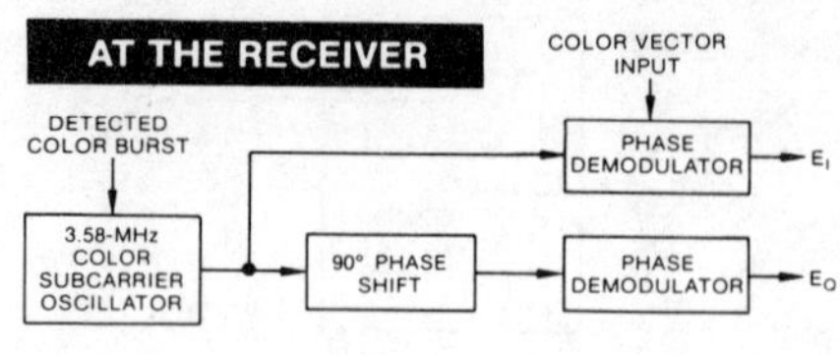

1. COLOR TV CHROMINANCE INFORMATION is transmitted on a subcarrier whose frequency is 3.579545 MHz. It is recovered at the receiver by developing two signals of that frequency in quadrature.

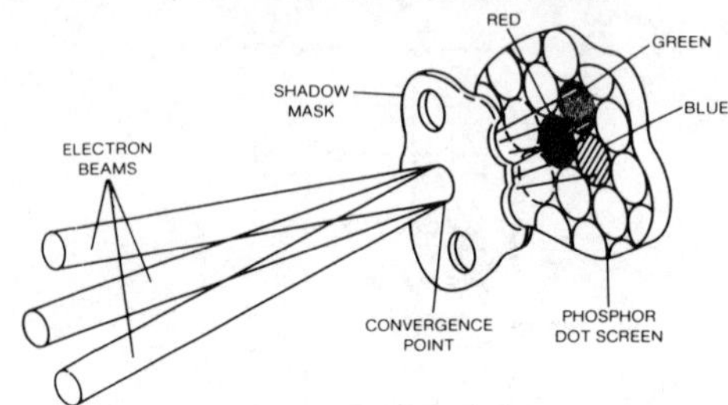

2. THE COLOR PICTURE TUBE (common version) has three electron guns. The three beams are controlled so that each goes through a hole in a mask and strikes its own phosphor color dot.

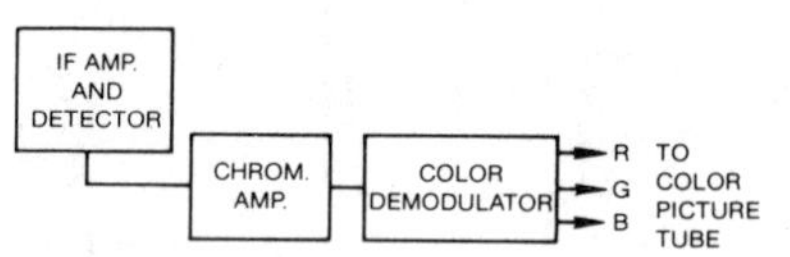

3. THE COLOR PROCESSING CIRCUITS receive the chrominance sidebands and burst signal from the video detector, demodulate in proper phases, and produce color signals for the CRT.

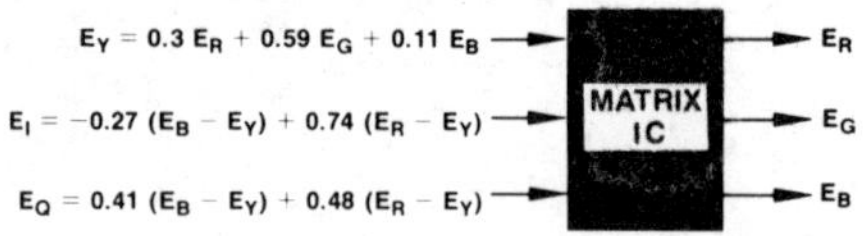

4. THE COLOR SIGNALS for the picture tube are obtained by matrixing the demodulated E_I, E_Q, and Y signals in proper proportion.

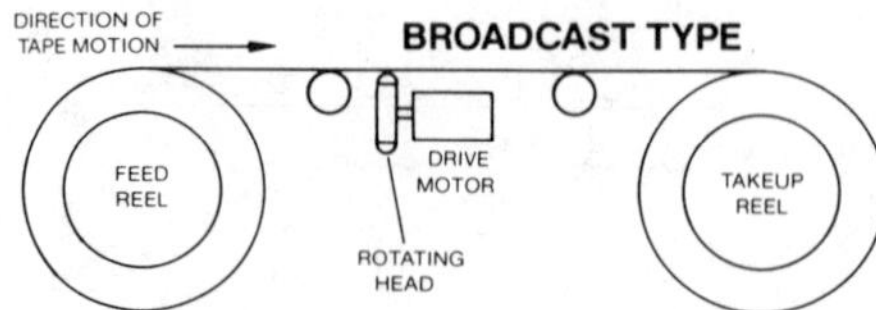

5. VIDEO TAPE RECORDERS became more practical as a result of the development of *helical scan* and *rotating heads*.

Self-Test—Review Questions

1. What is the difference between a black-and-white and color TV signal? Give details.
2. Define the terms luminance and chrominance.
3. Describe how a color image is formed.
4. What are the basic differences between a black-and-white TV picture tube and a color TV picture tube? Why are these differences significant?
5. Are there any significant differences between the RF/IF video-sync portions of a black-and-white and color TV receiver? Explain.
6. What is the color subcarrier reference? Where is it found? How is it recovered? What is it needed for?
7. Are there significant differences in the horizontal- and vertical-sweep circuits in black-and-white and color TV receivers? Specify and explain.
8. Draw a block diagram of the elements of the color portion of a TV receiver. Explain the function of each element. Describe how the chrominance signals are recovered.
9. Why are elaborate convergence and similar beam-control adjustments necessary in a color TV receiver?
10. Explain the advantages and disadvantages of video tape recording compared to video disk recording. For home use, what capability do tape machines have that disks do not have?

Learning Objectives—Next Volume

Overview—In Volume 5 of this series you will study the many ways digital signals are used, both in computers and other applications, and how computers work.

Epilogue

At this point in your studies, you have learned *how information is transformed* into electrical signals by appropriate transducers, used to modulate a suitable carrier, and radiated through space or conducted by wire to another point. The receiver at the remote point converts the received signal back to a reproduction of the original electrical signal that will *reproduce the original information* when supplied to an appropriate transducer. As you learned, the range of information that can be transferred to a remote point is almost limitless, ranging from simple on-off control signals to audio/video/digital data.

The important point is that the *basic elements* of transmission-reception systems are *always the same.* While the nature of the input and output transducers may vary, and the bandwidth of the signal depends on the information content of the signal, the system elements are the same. If you keep this concept in mind, you will be able to understand any system, no matter how complex it may seem at first glance. You will be able to separate it into its component parts, and these will be the system elements you have already studied. Until now, the information transferred was primarily in analog form rather than digital; it should be stressed, however, that this is *not* of fundamental importance and does not significantly change the system, because digital signals are not materially different from any other information.

In Volume 5, you will concentrate on the study of digital systems. As you will see, although the manipulation of information is accomplished in a different way, the objectives are consistent with what you have already learned. Since digital systems are becoming increasingly important, their study is an important part of our overall study. However, as you will see, all that you have already learned will apply to your study of digital systems and the management of information.

INDEX TO VOL. 4

(Note: A cumulative index covering all five volumes in this series will be found at the end of Volume 5.)